Advancing Maths for AQA

Revise for CORE 4

Sam Boardman   Tony Clough

Series editors
Sam Boardman   **Roger Williamson**   **Ted Graham**

www.heinemann.co.uk
✓ Free online support
✓ Useful weblinks
✓ 24 hour online ordering

01865 888058

Heinemann
Inspiring generations

Heinemann is an imprint of Pearson Education Limited,
a company incorporated in England and Wales, having its
registered office at Edinburgh Gate, Harlow, Essex, GM20 2JE.
Registered company number: 872828

Heinemann is the registered trademark of
Pearson Education Limited

First published 2006

13
10 9 8 7 6 5 4

British Library Cataloguing in Publication Data is available from the British
Library on request.

ISBN 978-0435513597

Typeset and illustrated by Tek-Art
Original illustrations © Tek-Art
Cover design by mccdesign ltd.
Printed in Malaysia, CTP-KHL

About this book

This book is designed to help you get your best possible grade in your Pure Core Maths 4 examination. The authors are Principal and Core examiners, and have a good understanding of AQA's requirements.

Revise for Core 4 covers the key topics that are tested in the Core 4 exam paper. You can use this book to help you revise at the end of your course, or you can use it throughout your course alongside the course textbook, *Advancing Maths for AQA AS & A level Pure Core Maths 3 & 4*, which provides complete coverage of the syllabus.

Helping you prepare for your exam

To help you prepare, each topic offers you:

- **Key points to remember** – summarise the mathematical ideas you need to know and be able to use.

- **Worked examples** – help you understand and remember important methods, and show you how to set out your answers clearly.

- **Revision exercises** – help you practise using these important methods to solve problems. Exam-level questions are included so you can be sure you are reaching the right standard, and answers are given at the back of the book so you can assess your progress.

- **Test Yourself questions** – help you see where you need extra revision and practice. If you do need extra help, they show you where to look in the *Advancing Maths for AQA AS & A level Pure Core Maths 3 & 4* textbook and which example to refer to in this book.

Exam practice and advice on revising

Examination style paper – this paper at the end of the book provides a set of questions of examination standard. It gives you an opportunity to practise taking a complete exam before you meet the real thing. The answers are given at the back of the book.

How to revise – for advice on revising before the exam, read the *How to revise* section on the next page.

How to revise using this book

Making the best use of your revision time

The topics in this book have been arranged in a logical sequence so you can work your way through them from beginning to end. But **how** you work on them depends on how much time there is between now and your examination.

If you have plenty of time before the exam then you can **work through each topic in turn**, covering the key points and worked examples before doing the revision exercises and Test Yourself questions.

If you are short of time then you can **work through the Test Yourself sections** first, to help you see which topics you need to do further work on.

However much time you have to revise, make sure you break your revision into short blocks of about 40 minutes, separated by five- or ten-minute breaks. Nobody can study effectively for hours without a break.

Using the Test Yourself sections

Each Test Yourself section provides a set of key questions. Try each question:

- If you can do it and get the correct answer, then move on to the next topic. Come back to this topic later to consolidate your knowledge and understanding by working through the key points, worked examples and revision exercises.

- If you cannot do the question, or get an incorrect answer or part answer, then work through the key points, worked examples and revision exercises before trying the Test Yourself questions again. If you need more help, the cross-references beside each Test Yourself question show you where to find relevant information in the *Advancing Maths for AQA AS & A level Pure Core Maths 3 & 4* textbook and which example in *Revise for C4* to refer to.

Reviewing the key points

Most of the key points are straightforward ideas that you can learn: try to understand each one. Imagine explaining each idea to a friend in your own words, and say it out loud as you do so. This is a better way of making the ideas stick than just reading them silently from the page.

As you work through the book, remember to go back over key points from earlier topics at least once a week. This will help you to remember them in the exam.

CHAPTER 1

Binomial series expansion

Key points to remember

1 Using the formula for the sum of an infinite geometric series, the expansion of $(1 - x)^{-1}$ when $|x| < 1$ is $1 + x + x^2 + x^3 + \ldots + x^n + \ldots$.

2 When $|x| < 1$, the binomial series expansion formula is
$$(1 + x)^n = 1 + nx + \frac{n \times (n - 1)}{1 \times 2} x^2 + \frac{n \times (n - 1) \times (n - 2)}{1 \times 2 \times 3} x^3 + \ldots$$

3 When n is a positive integer, the series will have a finite number of terms with the highest power being x^n.

4 When n is real but **not** a positive integer, the series will have an infinite number of terms.

5 The expansion of $(1 + bx)^n$, where b is a constant, is valid for $|bx| < 1$ or when $|x| < \frac{1}{|b|}$.

6 The expansion of $(a + x)^n$ can be found by writing $(a + x)^n = a^n \left(1 + \frac{x}{a}\right)^n$.

7 The expansion of $(a + x)^n$, where a is a positive constant, is valid for $|x| < a$.

Worked example 1

(a) Find the expansion of $(1 - 2x)^{-1}$ ascending powers of x up to and including the term in x^4.

(b) Hence find the coefficient of x^4 in the expansion of $(3 + x)(1 - 2x)^{-1}$.

(a) $(1 - 2x)^{-1} = \dfrac{1}{1 - 2x}$ is of the same form as the sum to infinity of a geometric series.

Expansion of $(1 - y)^{-1}$ is $1 + y + y^2 + y^3 + y^4 + \ldots$ •——— Using **1**

Hence $(1 - 2x)^{-1}$ has expansion $1 + 2x + (2x)^2 + (2x)^3 + (2x)^4 + \ldots$ •——— Putting $y = 2x$

Required expansion is $1 + 2x + 4x^2 + 8x^3 + 16x^4$.

(b) In the product $(3 + x)(1 - 2x)^{-1}$, terms involving x^4 arise from multiplying 3 by $16x^4$ and x by $8x^3$.

Therefore, the coefficient of x^4 is $48 + 8 = 56$.

Worked example 2

(a) Find the binomial expansion of $(1 - 3x)^{-2}$ up to and including the term in x^3.

(b) Hence find the coefficient of x^3 in the series expansion of
$\dfrac{4 - 7x}{(1 - 3x)^2}$.

(c) Find the range of values of x for which the series expansion
of $\dfrac{4 - 7x}{(1 - 3x)^2}$ is valid.

(a) $(1 - 3x)^{-2} = 1 + (-2)(-3x) + \dfrac{(-2)(-3)}{(1 \times 2)}(-3x)^2 +$

$\quad\quad\quad \dfrac{(-2)(-3)(-4)}{(1 \times 2 \times 3)}(-3x)^3 + \ldots$ Using **2**

$\quad\quad = 1 + 6x + 3(9x^2) - 4(-27x^3) + \ldots$

$\quad\quad = 1 + 6x + 27x^2 + 108x^3$ (up to term in x^3)

(b) $\dfrac{4 - 7x}{(1 - 3x)^2} = (4 - 7x)(1 - 3x)^{-2}$

$\quad\quad\quad = (4 - 7x)(1 + 6x + 27x^2 + 108x^3 + \ldots)$

Terms in x^3 arise from $4 \times 108x^3 - 7 \times 27x^3 = 243x^3$

Hence, the coefficient of x^3 is 243.

(c) Expansion of $(1 - 3x)^{-2}$ is valid when $|3x| < 1$.   Using **5**

Hence the series expansion of $\dfrac{4 - 7x}{(1 - 3x)^2}$ is valid for

$|x| < \dfrac{1}{3}$ or when $-\dfrac{1}{3} < x < \dfrac{1}{3}$.

Worked example 3

(a) Given that $|x| < 4$, show that the binomial expansion of

$\sqrt{4 + x}$, up to the term in x^3, is $2 + \dfrac{x}{4} - \dfrac{x^2}{64} + \dfrac{x^3}{512}$.

(b) By substituting $x = -\dfrac{1}{9}$ into the expansion of part (a), find

an approximation for $\sqrt{35}$, giving your answer to three
decimal places.

(a) You can write $\sqrt{4+x} = \sqrt{4}\sqrt{1+\dfrac{x}{4}} = 2\left(1+\dfrac{x}{4}\right)^{\frac{1}{2}}$.

Using **6**

$\left(1+\dfrac{x}{4}\right)^{\frac{1}{2}}$

$= 1 + \dfrac{1}{2}\left(\dfrac{x}{4}\right) + \dfrac{\left(\frac{1}{2}\right)\left(-\frac{1}{2}\right)}{1 \times 2}\left(\dfrac{x}{4}\right)^2 + \dfrac{\left(\frac{1}{2}\right)\left(-\frac{1}{2}\right)\left(-\frac{3}{2}\right)}{1 \times 2 \times 3}\left(\dfrac{x}{4}\right)^3 + \dots$

Using **2**

$= 1 + \dfrac{x}{8} - \dfrac{x^2}{128} + \dfrac{x^3}{1024}$

Hence $\sqrt{4+x} = 2\left(1+\dfrac{x}{4}\right)^{\frac{1}{2}} = 2 + \dfrac{x}{4} - \dfrac{x^2}{64} + \dfrac{x^3}{512}$.

(b) Substituting $x = -\dfrac{1}{9}$ into $\sqrt{4+x}$ gives $\sqrt{\dfrac{35}{9}} = \dfrac{\sqrt{35}}{3}$

Substituting $x = -\dfrac{1}{9}$ into $2 + \dfrac{x}{4} - \dfrac{x^2}{64} + \dfrac{x^3}{512}$ gives

$$1.9720266\dots$$

Hence $\dfrac{\sqrt{35}}{3} \approx 1.9720$.

Therefore $\sqrt{35} \approx 5.916$ (to 3 decimal places).

Worked example 4

(a) Given that $|x| < \dfrac{1}{2}$, find the first three non-zero terms in

the binomial expansion of $\sqrt{1-4x^2}$.

(b) Hence, by integrating these three terms of the series find
the approximate value of $\int_{0}^{0.1} \sqrt{1-4x^2}\ \mathrm{d}x$, giving your answer
to 5 significant figures.

(a) $\sqrt{1 - 4x^2} = (1 - 4x^2)^{\frac{1}{2}}$

$(1 - 4x^2)^{\frac{1}{2}} = 1 + \frac{1}{2}(-4x^2) + \frac{\left(\frac{1}{2}\right)\left(-\frac{1}{2}\right)}{1 \times 2}(-4x^2)^2 + \ldots$

Using **2**

$= 1 - 2x^2 - 2x^4$

(b) Approximating using the three terms above

$\int_0^{0.1} \sqrt{1 - 4x^2}\ dx = \int_0^{0.1}(1 - 2x^2 - 2x^4)\ dx$

$= \left[x - \frac{2}{3}x^3 - \frac{2}{5}x^5\right]_0^{0.1}$

$= \left[0.1 - \frac{2}{3} \times 0.001 - \frac{2}{5} \times 0.00001\right] - 0 \approx 0.09932933 \ldots$

Integral is approximately 0.099329 (to 5 significant figures).

REVISION EXERCISE 1

1 (a) Find the expansion of $(1 - 3x)^{-1}$ in ascending powers of x up to and including the term in x^3.

(b) Hence find the coefficient of x^3 in the expansion of $(4 - 5x)(1 - 3x)^{-1}$.

2 (a) Obtain the binomial expansion of $(1 + 3x)^{-2}$ in ascending powers of x up to and including the term in x^3.

(b) State the range of values of x for which the full expansion is valid.

3 (a) Find the binomial expansions of each of the following up to and including the term in x^3.

(i) $(1 - x)^{\frac{1}{3}}$ (ii) $\sqrt{16 + x}$ (iii) $\dfrac{16}{(2 + 3x)^2}$

(b) For each of the expansions in part (a), determine the range of values of x for which the full expansion is valid.

4 Given that $|x| < 1$, find the first four terms in the expansion of $(1 - x^3)^{-\frac{1}{3}}$.

Hence find an approximation for $\displaystyle\int_0^{0.3} \frac{1}{\sqrt[3]{(1 + x^3)}}\ dx$, giving your answer to 5 decimal places.

5 (a) Find the binomial expansion of $(1 + 2x)^{-3}$ in ascending powers of x up to the term in x^3. State the set of values for which the expansion is valid.

(b) Hence find the coefficient of x^3 in the expansion of

$$\frac{(2 - 5x)}{(1 + 2x)^3}.$$

6 (a) Obtain the binomial expansion of $(1 + x)^{-4}$ up to and including the term in x^2.

(b) (i) Hence obtain the binomial expansion of $\left(1 + \frac{3}{2}x\right)^{-4}$ up to and including the term in x^2.

(ii) Find the range of values of x for which the expansion in **(b)(i)** is valid.

(c) Given that x is small, show that

$$\left(\frac{4}{2 + 3x}\right)^4 \approx a + bx + cx^2, \text{ where } a, b \text{ and } c \text{ are integers.}$$

7 (a) Find the binomial expansion of $(1 + x)^{\frac{1}{4}}$ up to and including the term in x^2.

(b) (i) Given that $|x| < 3.2$, show that

$$(16 + 5x)^{\frac{1}{4}} \approx 2 + \frac{5}{32}x - \frac{75}{4096}x^2.$$

(ii) Hence show that the fourth root of 21 is approximately 2.14.

8 (a) Find the binomial expansion of $(1 + x)^{-\frac{1}{2}}$ up to the term in x^3.

(b) Hence, or otherwise, obtain the binomial approximation

of $\dfrac{1}{\sqrt{1 - 5x}}$ up to the term in x^3.

(c) By substituting $x = 0.01$ into your answer for part **(b)**, show that $\sqrt{95} \approx 9.7468$.

9 The coefficient of x^3 in the expansion of $(1 + ax)^{-3}$ is 80. Find the value of a and the coefficient of x^2 in the expansion.

10 (a) Show that $\dfrac{1}{2 - x} + \dfrac{2}{1 + 2x} \equiv \dfrac{5}{(2 - x)(1 + 2x)}.$

(b) Find the binomial expansions of **(i)** $(2 - x)^{-1}$ and **(ii)** $(1 + 2x)^{-1}$ up to and including the term in x^3.

(c) Use the previous results to find the binomial expansion of $\dfrac{5}{(2 - x)(1 + 2x)}$ up to and including the term in x^3.

(d) Determine the range of values of x for which the expansion in part **(c)** is valid.

Test yourself	**What to review**
	If your answer is incorrect:

1 Given that $|x| < \frac{1}{3}$, find the first five terms in ascending powers of x in the expansion of $(1 - 3x)^{-1}$.

Review Advancing Maths for AQA C3C4 pages 183–184.

2 Given that $|x| < 1$, use the binomial series expansion formula to find the first three terms in ascending powers of x in the expansion of $(1 + x)^{-3}$.

Review Advancing Maths for AQA C3C4 pages 185–188.

3 **(a)** Find the binomial expansion of $(1 + 4x)^{\frac{3}{4}}$ in ascending powers of x up to the term in x^3.
(b) State the range of values of x for which the expansion in part **(a)** is valid.

Review Advancing Maths for AQA C3C4 pages 185–188.

4 Given that $|x| < 4$, write down the binomial expansion of $(4 + x)^{-\frac{1}{2}}$ in ascending powers of x up to and including the term in x^2.

Review Advancing Maths for AQA C3C4 pages 189–190.

5 Given that $|x| < 1$, use the binomial series expansion formula to find the first three non-zero terms in ascending powers of x in the expansion of $(1 + x^2)^{-\frac{1}{2}}$.
Use these three terms to find an approximation for
$$\int_0^{0.1} \frac{1}{\sqrt{(1 + x^2)}} \, dx$$
giving your answer to 3 significant figures.

Review Advancing Maths for AQA C3C4 pages 191–192.

6 Given that $|x| < \frac{1}{3}$, write down the binomial expansion of $(1 - 3x)^{-\frac{2}{3}}$ in ascending powers of x up to and including the term in x^3.
Hence obtain the coefficient of x^3 in the expansion of
$$\frac{2 + x}{\sqrt[3]{(1 - 3x)^2}}.$$

Review Heinemann Book C3C4 pages 191–197.

Test yourself   ANSWERS

1 $1 + 3x + 9x^2 + 27x^3 + 81x^4$

2 $1 - 3x + 6x^2$

3 **(a)** $1 + 3x - \frac{3}{2}x^2 + \frac{5}{2}x^3$ **(b)** $|x| < \frac{1}{4}$

4 $\frac{1}{2} - \frac{1}{16}x + \frac{3}{256}x^2$

5 $1 - \frac{1}{2}x^2 + \frac{3}{8}x^4$, 0.0998

6 $1 + 2x + 5x^2 + \frac{40}{3}x^3$, $\frac{95}{3}$

CHAPTER 2

Rational functions and division of polynomials

Key points to remember

1 To simplify rational expressions:
- factorise all algebraic expressions
- cancel any factors that are common to the numerator and denominator.

2 To multiply rational expressions:
- factorise all algebraic expressions
- write as a single fraction
- cancel any factors that are common to the numerator and denominator.

3 To divide by a rational expression:
- change the division to a multiplication of the reciprocal
- factorise all algebraic expressions
- write as a single fraction
- cancel any factors that are common to the numerator and denominator.

4 To add/subtract rational expressions:
- factorise all algebraic expressions
- write each rational expression with the same common denominator
- add/subtract to get a single rational expression
- simplify the numerator
- cancel any factors that are common to the numerator and denominator.

5 Polynomial $\equiv$ divisor $\times$ quotient + remainder

6 When a polynomial is divided by a linear expression the remainder will always be a constant and the quotient will always be one degree less than the polynomial.

7 Before starting to use algebraic long division write the polynomial and divisor in descending powers of x and include all powers of x in the polynomial, inserting zero coefficients if necessary.

8 A more general form of the **factor theorem** is

$(ax + b)$ is a factor of the polynomial $P(x) \Leftrightarrow P\left(-\dfrac{b}{a}\right) = 0$

9 **The remainder theorem:**

If a polynomial $P(x)$ is divided by $(ax + b)$, the remainder is $P\left(-\dfrac{b}{a}\right)$.

Worked example 1

Write $\dfrac{x+3}{x^2+3x+2} + \dfrac{2x}{x^2-4}$ as a single algebraic fraction in its simplest form.

2

$$\dfrac{x+3}{x^2+3x+2} + \dfrac{2x}{x^2-4}$$

	Follow the steps in **4**

$$= \dfrac{x+3}{(x+1)(x+2)} + \dfrac{2x}{(x+2)(x-2)}$$

Factorising the denominators

$$= \dfrac{(x+3)(x-2)}{(x+1)(x+2)(x-2)} + \dfrac{2x(x+1)}{(x+1)(x+2)(x-2)}$$

LCD is $(x+1)(x+2)(x-2)$

$$= \dfrac{x^2 - 2x + 3x - 6 + 2x^2 + 2x}{(x+1)(x+2)(x-2)}$$

$$= \dfrac{3x^2 + 3x - 6}{(x+1)(x+2)(x-2)}$$

$$= \dfrac{3(x^2+x-2)}{(x+1)(x+2)(x-2)} = \dfrac{3(x+2)(x-1)}{(x+1)(x+2)(x-2)}$$

Factorising the numerator

$$= \dfrac{3(x-1)}{(x+1)(x-2)}$$

Cancelling the common factor $(x+2)$

Worked example 2

The polynomials $f(x)$ and $g(x)$ are defined by $f(x) = 6x^2 - x - 2$ and $g(x) = 9x^3 - 4x$.

(a) By considering the values of $f\left(\dfrac{2}{3}\right)$ and $g\left(\dfrac{2}{3}\right)$, show that $f(x)$ and $g(x)$ have a common linear factor.

(b) Hence write $\dfrac{6x^2 - x - 2}{9x^3 - 4x}$ as a simplified algebraic fraction.

(a) $f\left(\dfrac{2}{3}\right) = 6 \times \left(\dfrac{2}{3}\right)^2 - \dfrac{2}{3} - 2 = \dfrac{24}{9} - \dfrac{2}{3} - 2 = 0$

$g\left(\dfrac{2}{3}\right) = 9 \times \left(\dfrac{2}{3}\right)^3 - 4 \times \dfrac{2}{3} = \dfrac{72}{27} - \dfrac{8}{3} = 0$

$f\left(\dfrac{2}{3}\right) = 0 \Rightarrow (3x-2)$ is a factor of $f(x)$

Using **8** with $a = 3$ and $b = -2$

$g\left(\dfrac{2}{3}\right) = 0 \Rightarrow (3x - 2)$ is a factor of $g(x)$

So $(3x - 2)$ is a common linear factor of $f(x)$ and $g(x)$.

(b) $\quad f(x) = 6x^2 - x - 2 = (3x - 2)(2x + 1)$

and $g(x) = 9x^3 - 4x = x(9x^2 - 4) = x(3x - 2)(3x + 2)$

> Using the steps in **1**

$$\Rightarrow \quad \frac{6x^2 - x - 2}{9x^3 - 4x} = \frac{(3x - 2)(2x + 1)}{x(3x - 2)(3x + 2)}$$

$$= \frac{(2x + 1)}{x(3x + 2)}$$

> Cancelling the common factor $(3x - 2)$

> $\dfrac{(2x + 1)}{x(3x + 2)}$ cannot be simplified further

Worked example 3

(a) Find the remainder when $4x^3 - x^2 + 12x - 1$ is divided by $(4x - 1)$.

(b) Given that $\dfrac{4x^3 - x^2 + 12x - 1}{4x - 1} \equiv x^2 + A + \dfrac{B}{4x - 1}$,

find the values of A and B.

(a) Let $P(x) = 4x^3 - x^2 + 12x - 1$

$\quad$ Remainder $= P\left(\dfrac{1}{4}\right) = 4 \times \left(\dfrac{1}{4}\right)^3 - \left(\dfrac{1}{4}\right)^2 + 12 \times \dfrac{1}{4} - 1$

> Using **9** with $a = 4$ and $b = -1$

$\quad$ Remainder $= 2$

(b) $\quad \dfrac{4x^3 - x^2 + 12x - 1}{4x - 1} \equiv x^2 + A + \dfrac{B}{4x - 1}$

$\Rightarrow 4x^3 - x^2 + 12x - 1 \equiv (4x - 1)(x^2 + A) + B$

> Multiplying both sides by $(4x - 1)$

$\Rightarrow B = $ remainder $\Rightarrow B = 2$

$\quad 4x^3 - x^2 + 12x - 1 \equiv (4x - 1)(x^2 + A) + 2$

> Using **5** with divisor $(4x - 1)$ and answer to **(a)**

$\Rightarrow 4x^3 - x^2 + 12x - 1 \equiv 4x^3 - x^2 + 4Ax - A + 2$

Comparing the x terms: $12x \equiv 4Ax \Rightarrow A = 3$

> There are many other ways to find A, eg. put $x = 0$ to get $-1 = -A + 2 \Rightarrow A = 3$

> You could use the method of long division (Using **7**) to answer **worked example 3**.
>
> $$\require{enclose}\begin{array}{r} x^2 + 3 \\ 4x - 1 \enclose{longdiv}{4x^3 - x^2 + 12x - 1} \\ \underline{4x^3 - x^2 \qquad\quad} \\ 0 + 12x - 1 \\ \underline{+ 12x - 3} \\ + 2 \end{array}$$
>
> $x^2 + 3$ is the quotient and 2 is the remainder.

Worked example 4

The polynomial P(x) is defined by P(x) = $4x^3 - 12x^2 + kx + 6$,
where k is a constant.
When P(x) is divided by $(2x - 1)$ the remainder is 6.

(a) Show that $k = 5$.

(b) **(i)** Hence find the value of P$\left(-\dfrac{1}{2}\right)$.

   (ii) Write P(x) as a product of three linear factors.

(c) Hence show that $\dfrac{2x^2 - 3x - 2}{4x^3 - 12x^2 + 5x + 6}$ can be simplified to

$\dfrac{1}{px + q}$, where p and q are integers.

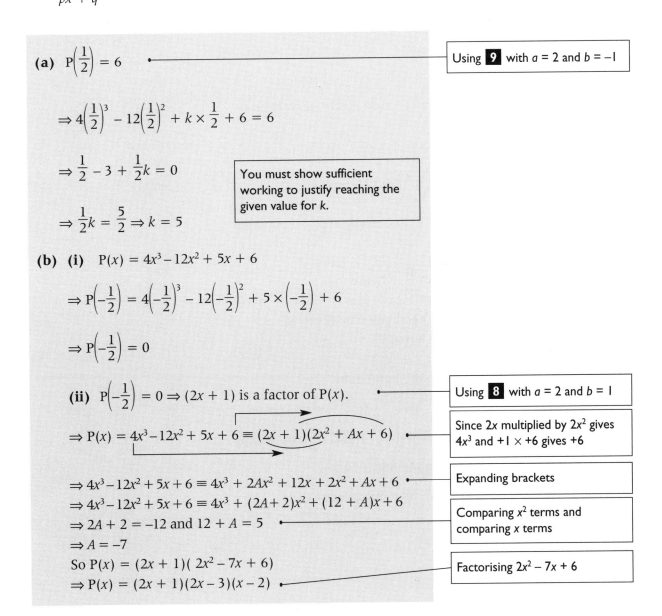

(a) P$\left(\dfrac{1}{2}\right) = 6$ — Using **9** with $a = 2$ and $b = -1$

$\Rightarrow 4\left(\dfrac{1}{2}\right)^3 - 12\left(\dfrac{1}{2}\right)^2 + k \times \dfrac{1}{2} + 6 = 6$

$\Rightarrow \dfrac{1}{2} - 3 + \dfrac{1}{2}k = 0$

> You must show sufficient working to justify reaching the given value for k.

$\Rightarrow \dfrac{1}{2}k = \dfrac{5}{2} \Rightarrow k = 5$

(b) **(i)** P(x) = $4x^3 - 12x^2 + 5x + 6$

$\Rightarrow$ P$\left(-\dfrac{1}{2}\right) = 4\left(-\dfrac{1}{2}\right)^3 - 12\left(-\dfrac{1}{2}\right)^2 + 5 \times \left(-\dfrac{1}{2}\right) + 6$

$\Rightarrow$ P$\left(-\dfrac{1}{2}\right) = 0$

   (ii) P$\left(-\dfrac{1}{2}\right) = 0 \Rightarrow (2x + 1)$ is a factor of P(x). — Using **8** with $a = 2$ and $b = 1$

$\Rightarrow$ P(x) = $4x^3 - 12x^2 + 5x + 6 \equiv (2x + 1)(2x^2 + Ax + 6)$ — Since $2x$ multiplied by $2x^2$ gives $4x^3$ and $+1 \times +6$ gives $+6$

$\Rightarrow 4x^3 - 12x^2 + 5x + 6 \equiv 4x^3 + 2Ax^2 + 12x + 2x^2 + Ax + 6$ — Expanding brackets

$\Rightarrow 4x^3 - 12x^2 + 5x + 6 \equiv 4x^3 + (2A + 2)x^2 + (12 + A)x + 6$

$\Rightarrow 2A + 2 = -12$ and $12 + A = 5$ — Comparing x^2 terms and comparing x terms

$\Rightarrow A = -7$

So P(x) = $(2x + 1)(2x^2 - 7x + 6)$ — Factorising $2x^2 - 7x + 6$

$\Rightarrow$ P(x) = $(2x + 1)(2x - 3)(x - 2)$

(c) $\dfrac{2x^2 - 3x - 2}{4x^3 - 12x^2 + 5x + 6} \equiv \dfrac{(2x + 1)(x - 2)}{(2x + 1)(2x - 3)(x - 2)}$ ◂━━━ Using **I** and answer to part **(b)**

$\dfrac{2x^2 - 3x - 2}{4x^3 - 12x^2 + 5x + 6} \equiv \dfrac{1}{2x - 3}$ ◂━━━ Cancelling the common factors $(2x + 2)$ and $(x - 2)$

REVISION EXERCISE 2

1 Simplify

(a) $\dfrac{2x^2 + 8x}{x^2 + 3x - 4}$

(b) $\dfrac{x^2 - 4}{x^2 - 5x + 6} \times \dfrac{x^2 - 2x - 3}{x^2 - 1}$

(c) $\dfrac{2x^2 - 18}{x^2 - 3x} \div \dfrac{4x^2 + 12x}{2x^3 + 8x^2}$

2 Write $\dfrac{9}{x^2 + 5x + 4} + \dfrac{3}{x + 4}$ as a single algebraic fraction in its simplest form.

3 It is given that $y = \dfrac{4}{x - 1} - \dfrac{5x - 13}{(x - 1)(x - 3)}$.

(a) Write $\dfrac{4}{x - 1} - \dfrac{5x - 13}{(x - 1)(x - 3)}$ as a single algebraic fraction in its simplest form.

(b) Hence, or otherwise, find $\dfrac{dy}{dx}$ when $x = 2$.

4 The polynomial $(3 + x^2 + 6x^3)$ is divided by $(2x + 3)$.
Find **(a)** the quotient **(b)** the remainder.

5 The polynomials $f(x)$ and $g(x)$ are defined by
$f(x) = 2x^3 + x^2 - 8x - 4$ and $g(x) = 2x^2 + 9x + 4$.

(a) By considering the values of $f\left(-\dfrac{1}{2}\right)$ and $g\left(-\dfrac{1}{2}\right)$, show that $f(x)$ and $g(x)$ have a common linear factor.

(b) Write $f(x)$ as a product of three linear factors.

(c) Hence simplify $\dfrac{f(x)}{g(x)}$.

6 (a) Find the remainder when $4x^3 + 10x^2 - 4x + 37$ is divided by $(2x + 7)$.

(b) Given that
$$\frac{4x^3 + 10x^2 - 4x + 37}{2x + 7} \equiv 2x^2 + Ax + B + \frac{C}{2x + 7},$$
find the values of A, B and C.

(c) Hence find $\displaystyle\int_0^3 \frac{4x^3 + 10x^2 - 4x + 37}{2x + 7}\, dx$, giving your answer in an exact form.

7 When the polynomial $P(x) = 9x^3 - 6x^2 + kx + 5$, where k is a constant, is divided by $(3x - 1)$ the remainder is 4.

(a) Find the value of k.

(b) Hence find the remainder when $P(x)$ is divided by $(x + 1)$.

8 It is given that $P(x) = 6x^3 - x^2 + 10x - 8$.

(a) Use the factor theorem to show that $(3x - 2)$ is a factor of $P(x)$.

(b) Write $P(x)$ in the form $(3x - 2)(ax^2 + bx + c)$, where a, b and c are integers.

(c) Hence prove that the equation $P(x) = 0$ has only one real root.

9 (a) Given that $(2x + 1)$ is a factor of $f(x) = 2x^3 - 7x^2 + px + q$, where p and q are constants, show that $p = 2q - 4$.

(b) Given that $(x - 1)$ is also a factor of $f(x)$, find the values of p and q.

(c) Write $f(x)$ as a product of three linear factors.

(d) Hence, or otherwise, solve the equation $f(e^x) = 0$.

10 The polynomials $f(x)$ and $g(x)$ are defined by
$$f(x) = 4x^3 + px^2 + x - 14, \quad g(x) = 8x^3 - 8x^2 + px - 5,$$
where p is a constant.

When $f(x)$ and $g(x)$ are divided by $(2x - 3)$, the remainder is R in each case.

(a) Show that $p = 4$ and find the value of R.

(b) Given that $f(x) \equiv (2x - 3)Q(x) + R$ find the quadratic $Q(x)$.

(c) Write the cubic expression $[g(x) - f(x)]$ in the form $(ax + b)(cx^2 + dx + e)$, where a, b, c, d and e are integers.

2

Test yourself	**What to review**

1 Simplify $\dfrac{x-2}{x^2+4x+3} \div \dfrac{x^2-4}{x^2+3x+2}$.

Review Advancing Maths for
AQA C3C4 pages 198–201.

---

2 Write $\dfrac{x-16}{x^2+4x} - \dfrac{5}{x+4}$ as a single algebraic fraction
in its simplest form.

Review Advancing Maths for
AQA C3C4 pages 201–203.

---

3 It is given that $P(x) = 4x^3 - 12x^2 + 3x + 5$.

Review Advancing Maths for
AQA C3C4 pages 206–209.

   (a) Find the value of $P\left(\dfrac{5}{2}\right)$.

   (b) Write $P(x)$ as a product of three linear factors.

---

4 Find the quotient and remainder when the polynomial
$(6x^3 + 5x^2 + 4)$ is divided by $(1 + 2x)$.

Review Advancing Maths for
AQA C3C4 pages 203–206.

---

5 When the polynomial $P(x) = 4x^3 + kx^2 - 19x - 7$, where k is
a constant, is divided by $(2x + 3)$ the remainder is -10.

Review Advancing Maths for
AQA C3C4 pages 206–212.

   (a) Find the value of k.

   (b) Hence, use the factor theorem to show that $(2x + 1)$ is
a factor of $P(x)$.

   (c) Find the roots of the equation $P(x) = 0$.

Test yourself ANSWERS

5 (a) $k = -8$ **(c)** $-1, -0.5, 3.5$

4 Quotient $= 3x^2 + x - \dfrac{1}{2}$,   Remainder $= 4.5$

3 (a) 0 **(b)** $(2x - 5)(2x + 1)(x - 1)$

2 $-\dfrac{4}{x}$

1 $\dfrac{1}{x+3}$

CHAPTER 3

Partial fractions and applications

Key points to remember

1   Only proper fractions can be expressed in terms of partial fractions. If the given rational expression is improper you must first carry out a long division to obtain a proper fraction.

2   A proper fraction which has up to three linear factors in the denominator, $\dfrac{p(x)}{(x-a)(x-b)(x-c)}$, has three partial fractions of the form

$$\dfrac{A}{(x-a)} + \dfrac{B}{(x-b)} + \dfrac{C}{(x-c)}.$$

3   A proper fraction which has a repeated linear factor in the denominator, $\dfrac{q(x)}{(x-a)^3}$, has partial fractions of the form $\dfrac{A}{(x-a)} + \dfrac{B}{(x-a)^2} + \dfrac{C}{(x-a)^3}$.

4   To solve integration problems using partial fractions, the following two integrals, where a and b are constants, are often needed:

$$\int \frac{1}{ax+b}\,dx = \frac{1}{a}\ln|ax+b| + c$$

$$\int \frac{1}{(ax+b)^2}\,dx = -\frac{1}{a(ax+b)} + c$$

5   You will frequently need to use the following two applications (for $n = -1$ and $n = -2$) of the binomial expansion:

$$(1+y)^{-1} = 1 - y + y^2 - y^3 + y^4 - \ldots \text{ valid for } |y| < 1$$

$$(1+y)^{-2} = 1 - 2y + 3y^2 - 4y^3 + \ldots \text{ valid for } |y| < 1$$

Worked example 1

(a)   Express $\dfrac{5x}{(x+2)(2x-1)}$ in partial fractions.

(b)   Hence find $\displaystyle\int \frac{5x}{(x+2)(2x-1)}\,dx$.

(a) $\dfrac{5x}{(x+2)(2x-1)} \equiv \dfrac{A}{(x+2)} + \dfrac{B}{(2x-1)}$

Using **2**

$\dfrac{5x}{(x+2)(2x-1)} \equiv \dfrac{A(2x-1) + B(x+2)}{(x+2)(2x-1)}$

Writing both sides with the same denominator.

$\Rightarrow 5x \equiv A(2x-1) + B(x+2)$

Equating the numerators since denominators are equal.

Let $x = -2 \Rightarrow -10 = -5A \Rightarrow A = 2$

Let $x = \dfrac{1}{2} \Rightarrow 2.5 = 2.5B \Rightarrow B = 1$

Choosing values of x which lead to each bracket being zero.

$\dfrac{5x}{(x+2)(2x-1)} \equiv \dfrac{2}{x+2} + \dfrac{1}{2x-1}$

Check: equate coefficients of x
$\Rightarrow 5 \equiv 2A + B, \quad 5 = 2 \times 2 + 1$
which is true.

(b) $\displaystyle\int \dfrac{5x}{(x+2)(2x-1)}\, dx = \int \left(\dfrac{2}{x+2} + \dfrac{1}{2x-1} \right) dx$

Using part **(a)**

$= \displaystyle\int \dfrac{2}{x+2}\, dx + \int \dfrac{1}{2x-1}\, dx$

Splitting the integrals

$= 2\displaystyle\int \dfrac{1}{x+2}\, dx + \dfrac{1}{2}\int \dfrac{2}{2x-1}\, dx$

Writing in the form $\displaystyle\int \dfrac{f'(x)}{f(x)}\, dx$

$= 2\ln|x+2| + \dfrac{1}{2}\ln|2x-1| + c$

Using **4**

Worked example 2

(a) Given that $\dfrac{2x^2 + 7x + 8}{x(x+2)^2} \equiv \dfrac{A}{x} + \dfrac{B}{x+2} + \dfrac{C}{(x+2)^2}$ find

the values of A and C and show that $B = 0$.

This form comes from using **2** and **3**

(b) Hence show that $\displaystyle\int_{2}^{6} \dfrac{2x^2 + 7x + 8}{x(x+2)^2}\, dx = 2\ln3 - \dfrac{1}{8}$.

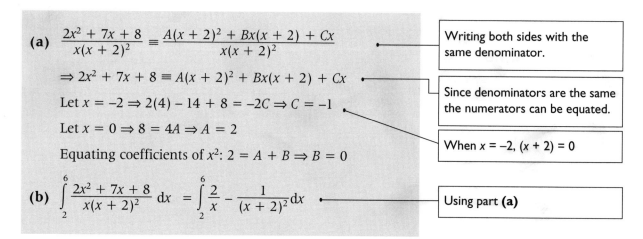

(a) $\dfrac{2x^2 + 7x + 8}{x(x+2)^2} \equiv \dfrac{A(x+2)^2 + Bx(x+2) + Cx}{x(x+2)^2}$

Writing both sides with the same denominator.

$\Rightarrow 2x^2 + 7x + 8 \equiv A(x+2)^2 + Bx(x+2) + Cx$

Since denominators are the same the numerators can be equated.

Let $x = -2 \Rightarrow 2(4) - 14 + 8 = -2C \Rightarrow C = -1$

Let $x = 0 \Rightarrow 8 = 4A \Rightarrow A = 2$

When $x = -2$, $(x+2) = 0$

Equating coefficients of x^2: $2 = A + B \Rightarrow B = 0$

(b) $\displaystyle\int_{2}^{6} \dfrac{2x^2 + 7x + 8}{x(x+2)^2}\, dx = \int_{2}^{6} \dfrac{2}{x} - \dfrac{1}{(x+2)^2}\, dx$

Using part **(a)**

$$= \left[2\ln x + \frac{1}{(x + 2)} \right]_2^6$$

Using **4**

$$= \left(2\ln 6 + \frac{1}{8} \right) - \left(2\ln 2 + \frac{1}{4} \right)$$

$$= 2\ln \frac{6}{2} - \frac{1}{8}$$

Using $\ln a - \ln b = \ln \dfrac{a}{b}$

$$= 2\ln 3 - \frac{1}{8}$$

3

Worked example 3

(a) Given that $\dfrac{x^2 + 2x + 2}{(1 + x)^2(2 + x)} \equiv \dfrac{A}{1 + x} + \dfrac{B}{(1 + x)^2} + \dfrac{C}{2 + x}$

find the values of A, B and C.

(b) Hence find the binomial expansion of $\dfrac{x^2 + 2x + 2}{(1 + x)^2(2 + x)}$

up to and including the term in x^2, giving each term in its simplest form.

(a) $\dfrac{x^2 + 2x + 2}{(1 + x)^2(2 + x)} \equiv \dfrac{A(1 + x)(2 + x) + B(2 + x) + C(1 + x)^2}{(1 + x)^2(2 + x)}$

Writing both sides with the same denominator

$\Rightarrow x^2 + 2x + 2 \equiv A(1 + x)(2 + x) + B(2 + x) + C(1 + x)^2$

Let $x = -2 \Rightarrow 4 - 4 + 2 = C \Rightarrow C = 2$

Let $x = -1 \Rightarrow 1 - 2 + 2 = B \Rightarrow B = 1$

Equating coefficients of x^2: $1 = A + C \Rightarrow A = -1$

Check: Put $x = 0$
$\Rightarrow 2 = A(1)(2) + B(2) + C(1)^2$
$2 = -1 \times 2 + 1 \times 2 + 2 \times 1$
which is true.

(b) $\dfrac{x^2 + 2x + 2}{(1 + x)^2(2 + x)} = \dfrac{-1}{1 + x} + \dfrac{1}{(1 + x)^2} + \dfrac{2}{2 + x}$

Using part **(a)**

$$= \frac{-1}{1 + x} + \frac{1}{(1 + x)^2} + \frac{2}{2\left(1 + \dfrac{x}{2} \right)}$$

$$= -1(1 + x)^{-1} + 1(1 + x)^{-2} + \left(1 + \frac{x}{2} \right)^{-1}$$

Writing in the form $(1 + y)^n$.

$$= -1(1 - x + x^2 - \ldots) + 1(1 - 2x + 3x^2 - \ldots) +$$

$$\left[1 - \left(\frac{x}{2} \right) + \left(\frac{x}{2} \right)^2 - \ldots \right]$$

Using **5**

Use brackets around the $\dfrac{x}{2}$. A common error is to not square the 2.

$$= (-1 + 1 + 1) + (x - 2x - \frac{1}{2}x) + (-x^2 + 3x^2 + \frac{1}{4}x^2) + \dots$$

$$\frac{x^2 + 2x + 2}{(1 + x)^2(2 + x)} = 1 - \frac{3}{2}x + \frac{9}{4}x^2 + \dots$$

> **Check:** Both sides equal 1 when $x = 0$.

Worked example 4

(a) Express $\dfrac{8x^2 - 23x - 18}{(2x + 1)(x - 4)}$ in the form $A + \dfrac{B}{2x + 1} + \dfrac{C}{x - 4}$.

> This form comes from using
> **1** and **2**

(b) It is given that $y = \dfrac{8x^2 - 23x - 18}{(2x + 1)(x - 4)}$.

Find the value of $\dfrac{d^2y}{dx^2}$ when $x = 0$.

(a) $\dfrac{8x^2 - 23x - 18}{(2x + 1)(x - 4)} \equiv$

$$\frac{A(2x + 1)(x - 4) + B(x - 4) + C(2x + 1)}{(2x + 1)(x - 4)}$$

> **Check:** Put $x = 0$
> $\Rightarrow 18 = -4A - 4B + C$
> $-18 = -16 - 4 + 2$
> which is true.

$$\Rightarrow 8x^2 - 23x - 18 \equiv A(2x + 1)(x - 4) + B(x - 4) + C(2x + 1)$$

Equating coefficients of x^2: $8 = 2A \Rightarrow A = 4$

Let $x = 4$ $\Rightarrow 8(16) - 23(4) - 18 = C(9) \Rightarrow 9C = 18 \Rightarrow C = 2$

Let $x = -0.5 \Rightarrow 2 + 11.5 - 18 = -4.5B \Rightarrow -4.5B = -4.5 \Rightarrow B = 1$

(b) $y = 4 + \dfrac{1}{2x + 1} + \dfrac{2}{x - 4}$

> Using part **(a)**

$y = 4 + (2x + 1)^{-1} + 2(x - 4)^{-1}$

> Writing in the form $(ax + b)^n$.

$\dfrac{dy}{dx} = -1(2x + 1)^{-2}(2) - 2(x - 4)^{-2}(1)$

$\dfrac{dy}{dx} = -2(2x + 1)^{-2} - 2(x - 4)^{-2}$

> Using $\dfrac{d}{dx}\left((ax + b)^n\right)$
> $= n(ax + b)^{n-1}(a)$

$\dfrac{d^2y}{dx^2} = 4(2x + 1)^{-3}(2) + 4(x - 4)^{-3}(1)$

When $x = 0$, $\dfrac{d^2y}{dx^2} = 8(1)^{-3} + 4(-4)^{-3} = 8 - \dfrac{4}{64} = 7\dfrac{15}{16}$

REVISION EXERCISE 3

1 (a) Show that $\dfrac{2x-5}{x+1}$ can be written in the form

$A + \dfrac{B}{x+1}$, where A and B are integers.

(b) Hence show that $\displaystyle\int_{1}^{3} \dfrac{2x-5}{x+1}\, dx = 4 - 7\ln 2$.

2 (a) Express $\dfrac{4x^2+1}{2x-1}$ in the form $Ax + B + \dfrac{C}{2x-1}$.

(b) Hence find $\displaystyle\int \dfrac{4x^2+1}{2x-1}\, dx$.

3 (a) Express $\dfrac{5x+7}{(x+2)(x-1)}$ as the sum of two partial fractions.

(b) Hence find the value of $\displaystyle\int_{2}^{3} \dfrac{5x+7}{(x+2)(x-1)}\, dx$, leaving your answer in the form $\ln k$ where k is an integer.

4 Show that $\displaystyle\int_{4}^{6} \dfrac{2}{x(x-2)}\, dx = \ln \dfrac{4}{3}$.

5 (a) Given that
$$\dfrac{x^2-3}{(x+1)^2(x+2)} \equiv \dfrac{A}{x+1} + \dfrac{B}{(x+1)^2} + \dfrac{C}{x+2},$$
find the values of B and C and show that $A = 0$.

(b) Hence show that $\displaystyle\int_{0}^{4} \dfrac{x^2-3}{(x+1)^2(x+2)}\, dx = \ln 3 - \dfrac{8}{5}$.

6 (a) Express $\dfrac{3x-10}{9x^2-4}$ in the form $\dfrac{A}{3x+2} + \dfrac{B}{3x-2}$.

(b) Hence find $\displaystyle\int \dfrac{3x-10}{9x^2-4}\, dx$.

7 (a) Given that $\dfrac{3-7x-4x^2}{(1+x)(1-2x)} \equiv A + \dfrac{B}{1+x} + \dfrac{C}{1-2x}$,
find the values of A, B and C.

(b) Hence find the binomial expansion of $\dfrac{3-7x-4x^2}{(1+x)(1-2x)}$
up to and including the term in x^3, giving each term in its simplest form.

8 (a) Given that
$$\frac{x^2 + 7x + 4}{(1 + x)^2(1 + 2x)} \equiv \frac{A}{1 + 2x} + \frac{B}{1 + x} + \frac{C}{(1 + x)^2},$$
find the values of A, B and C.

(b) Hence find the binomial expansion of $\dfrac{x^2 + 7x + 4}{(1 + x)^2(1 + 2x)}$
up to and including the term in x^2, giving each term in its simplest form.

(c) Find the range of values of x for which this expansion is valid.

9 (a) Express $\dfrac{6x^2 + 29x + 23}{(2x + 3)(x + 5)}$ in the form

$$A + \frac{B}{2x + 3} + \frac{C}{x + 5}.$$

(b) It is given that $y = \dfrac{6x^2 + 29x + 23}{(2x + 3)(x + 5)}$.

Find the value of $\dfrac{d^2y}{dx^2}$ when $x = -1$.

10 (a) Express $\dfrac{8x^3 - 18x^2}{(x - 1)(4x + 1)}$ in the form

$$Ax + B + \frac{C}{x - 1} + \frac{D}{4x + 1}.$$

(b) It is given that $y = \dfrac{8x^3 - 18x^2}{(x - 1)(4x + 1)}$.

Find the value of $\dfrac{d^2y}{dx^2}$ when $x = 0$.

Test yourself

What to review

If your answer is incorrect:

1 Express $\dfrac{8x + 2}{(x + 2)(2x - 3)}$ as the sum of two partial fractions.

Review Advancing Maths for AQA C3C4 pages 215–219.

2 (a) Given that $\dfrac{x^2 + x - 8}{x(x + 4)} \equiv A + \dfrac{B}{x} + \dfrac{C}{x + 4}$ find the values of A, B and C.

Review Advancing Maths for AQA C3C4 pages 215–222.

(b) Hence find $\displaystyle\int \frac{x^2 + x - 8}{x(x + 4)}\, dx$.

Test yourself *(continued)*	**What to review**

3 **(a)** Given that

$$\frac{x^2 + x + 1}{(x + 2)(x + 1)^2} \equiv \frac{A}{x + 2} + \frac{B}{x + 1} + \frac{C}{(x + 1)^2},$$

find the values of A, B and C.

Review Advancing Maths for AQA C3C4 pages 215–222.

(b) Hence show that $\displaystyle\int_0^1 \frac{x^2 + x + 1}{(x + 2)(x + 1)^2} \, dx = \ln\left(\frac{27}{32}\right) + \frac{1}{2}.$

3

4 **(a)** Write down the binomial expansion of $(1 + 2x)^{-2}$ in ascending powers of x up to and including x^2.

(b) Express $\dfrac{18x^2 + 19x + 2}{(2 + x)(1 + 2x)^2}$ in the form

Review Advancing Maths for AQA C3C4 pages 222–228.

$$\frac{A}{2 + x} + \frac{B}{1 + 2x} + \frac{C}{(1 + 2x)^2}.$$

(c) Hence find the binomial expansion of

$$\frac{18x^2 + 19x + 2}{(2 + x)(1 + 2x)^2} \text{ up to and including the term in } x^2.$$

Test yourself ANSWERS

(c) $1 + 5x - 19\tfrac{1}{2}x^2$

4 (a) $1 - 4x + 12x^2$ **(b)** $\dfrac{4}{2 + x} + \dfrac{1}{1 + 2x} - \dfrac{2}{(1 + 2x)^2}$

3 (a) $A = 3$, $B = -2$, $C = 1$

2 (a) $A = 1$, $B = -2$, $C = -1$ **(b)** $x - 2\ln|x| - \ln|x + 4| + c$

1 $\dfrac{2}{x + 2} + \dfrac{4}{2x - 3}$

CHAPTER 4

Implicit differentiation and applications

Key points to remember

1 y is an **implicit function** of x if y cannot be written in the form $y = f(x)$; for example the equation $y^2 + x\ln y = x + \sin y$.

2
$$\frac{d}{dx}(y^n) = ny^{n-1}\frac{dy}{dx}$$

$$\frac{d}{dx}(\ln y) = \frac{1}{y}\frac{dy}{dx}$$

$$\frac{d}{dx}(\sin y) = \cos y \frac{dy}{dx}$$

In general,

$$\frac{d}{dx}[f(y)] = f'(y)\frac{dy}{dx}, \text{ where } f'(y) = \frac{df}{dy}.$$

Worked example 1

A curve is defined by the equation $2y^3 - \ln y = x^4 - 7x$.

Find the value of $\dfrac{dy}{dx}$ at the point $(2, 1)$.

$$\frac{d}{dx}(2y^3 - \ln y) = \frac{d}{dx}(x^4 - 7x)$$
  — Differentiating both sides with respect to x

$$\frac{d}{dx}(2y^3) - \frac{d}{dx}(\ln y) = 4x^3 - 7$$

$$6y^2\frac{dy}{dx} - \frac{1}{y}\frac{dy}{dx} = 4x^3 - 7$$
  — Using **2**

At the point $(2, 1)$, $\quad 6\dfrac{dy}{dx} - 1\dfrac{dy}{dx} = 32 - 7$

$$\Rightarrow \frac{dy}{dx} = 5$$

Worked example 2

(a) Differentiate $e^{2x} y$ with respect to x.

(b) Given that $e^{2x} y - \cos y = x + y$, find $\dfrac{dy}{dx}$ in terms of x and y.

(a) $\dfrac{d}{dx}(e^{2x} y) = e^{2x} \dfrac{d}{dx}(y) + y \dfrac{d}{dx}(e^{2x})$

> $e^{2x} y$ is a product so using
> $\dfrac{d}{dx}(uv) = u\dfrac{dv}{dx} + v\dfrac{du}{dx}$.

$\qquad\qquad = e^{2x} \dfrac{dy}{dx} + y\,(2e^{2x})$

> Using $\dfrac{d}{dx}(e^{kx}) = ke^{kx}$.

$\dfrac{d}{dx}(e^{2x} y) = e^{2x} \dfrac{dy}{dx} + 2\,y\,e^{2x}$

4

(b) $\dfrac{d}{dx}(e^{2x} y - \cos y) = \dfrac{d}{dx}(x + y)$

$\Rightarrow e^{2x} \dfrac{dy}{dx} + 2y\,e^{2x} - \dfrac{d}{dx}(\cos y) = 1 + \dfrac{dy}{dx}$

> Using part **(a)**

$\Rightarrow e^{2x} \dfrac{dy}{dx} + 2y\,e^{2x} - (-\sin y)\dfrac{dy}{dx} = 1 + \dfrac{dy}{dx}$

> Using **2** with f(y) = $\cos y$

$\Rightarrow (e^{2x} + \sin y - 1)\dfrac{dy}{dx} = 1 - 2y\,e^{2x}$

$\Rightarrow \dfrac{dy}{dx} = \dfrac{1 - 2y\,e^{2x}}{e^{2x} + \sin y - 1}$

Worked example 3

A curve has equation $y^2 - xy + x^2 - 2x - 7 = 0$.

(a) Find $\dfrac{dy}{dx}$.

(b) Find the coordinates of the two stationary points of the curve.

(a) $2y\dfrac{dy}{dx} - [x\dfrac{dy}{dx} + (1)y] + 2x - 2 = 0$

> Differentiating using **2** and the product rule.

$2y\dfrac{dy}{dx} - x\dfrac{dy}{dx} - y + 2x - 2 = 0$

$(2y - x)\dfrac{dy}{dx} = y - 2x + 2$

$\Rightarrow \qquad \dfrac{dy}{dx} = \dfrac{y - 2x + 2}{2y - x}$

(b) $\dfrac{y - 2x + 2}{2y - x} = 0 \Rightarrow y - 2x + 2 = 0 \Rightarrow y = 2x - 2$

> At a stationary point, $\dfrac{dy}{dx} = 0$

On the curve, at the stationary points where $y = 2x - 2$,
$(2x - 2)^2 - x(2x - 2) + x^2 - 2x - 7 = 0$

> Substituting $y = 2x - 2$ in the equation of the curve.

$\Rightarrow 4x^2 - 8x + 4 - 2x^2 + 2x + x^2 - 2x - 7 = 0$
$\Rightarrow 3x^2 - 8x - 3 = 0$
$\Rightarrow (3x + 1)(x - 3) = 0 \Rightarrow x = -\dfrac{1}{3}, x = 3$

> The coordinates of the stationary points must satisfy both the equation of the curve and the equation $y = 2x - 2$.

When $x = -\dfrac{1}{3}$, $y = 2\left(-\dfrac{1}{3}\right) - 2 = -2\dfrac{2}{3}$

When $x = 3$, $y = 2(3) - 2 = 4$

> Substituting in $y = 2x - 2$ to find the y-coordinates of the stationary points.

The stationary points of the curve are $\left(-\dfrac{1}{3}, -2\dfrac{2}{3}\right)$ and $(3, 4)$.

Worked example 4

A curve has equation $y = (1 + x)^x$. Find an equation of the tangent to the curve at the point $(1, 2)$.

> Although y is given explicitly in terms of x, implicit differentiation is required in this example.

$y = (1 + x)^x \Rightarrow \ln y = \ln(1 + x)^x$

> Taking natural logarithms of both sides since the power involves x.

$\Rightarrow \ln y = x \ln(1 + x)$

> Using log law: $\log a^b = b \log a$.

$\dfrac{1}{y} \dfrac{dy}{dx} = x \times \left(\dfrac{1}{1 + x}\right) + \ln(1 + x) \times 1$

> Using **2** and the product rule.

At $(1, 2)$ $\qquad \dfrac{1}{2} \dfrac{dy}{dx} = \dfrac{1}{2} + \ln 2$

Gradient of tangent at $(1, 2)$ is $\dfrac{dy}{dx} = 1 + \ln 4$

> Using $2\ln 2 = \ln 2^2$

Equation of tangent at $(1, 2)$ is $y - 2 = (1 + \ln 4)(x - 1)$

> Using $y - y_1 = m(x - x_1)$

Worked example 5

The line $y = 1$ intersects the curve $2x^2 y = 3x + y^2 + 1$ at the points $(2, 1)$ and A.

(a) Find the coordinates of A.

(b) Find an equation of the normal to the curve at the point $(2, 1)$.

(a) At A, $y = 1$ and $2x^2 y = 3x + y^2 + 1$

$\Rightarrow 2x^2 = 3x + 2$ •—————————————— Substituting $y = 1$ into the equation of the curve.

$\Rightarrow 2x^2 - 3x - 2 = 0$

$\Rightarrow (2x + 1)(x - 2) = 0$ •—————————————— Factorising

$\Rightarrow x = -\dfrac{1}{2}, x = 2$

$\Rightarrow A$ has coordinates $\left(-\dfrac{1}{2}, 1\right)$.

(b) $\left(2x^2 \dfrac{dy}{dx} + 4xy\right) = 3 + 2y\dfrac{dy}{dx}$ •—————————————— Using **2** and the product rule

4

At $(2, 1)$ $8\dfrac{dy}{dx} + 8 = 3 + 2\dfrac{dy}{dx}$

Gradient of tangent at $(2, 1)$ is $\dfrac{dy}{dx} = -\dfrac{5}{6}$. •—————————————— Using $m_1 \times m_2 = -1$

$\Rightarrow$ Gradient of the normal at $(2, 1)$ is $\dfrac{6}{5}$. •—————————————— Using $y - y_1 = m(x - x_1)$

Equation of the normal at $(2, 1)$ is $y - 1 = \dfrac{6}{5}(x - 2)$ •—— Since the question asks for '**an equation**' any correct form of the equation of this normal would be acceptable in the examination.

or $5y = 6x - 7$.

REVISION EXERCISE 4

1 Differentiate the following with respect to x:

(a) $x + \ln y$ (b) $x^3 + y^3 + y - 7x$

(c) $x \sin 2y$ (d) $\dfrac{x}{y + 1} - 3$

2 A curve has equation $xy^2 + y = 6$. Use implicit differentiation to find the value of $\dfrac{dy}{dx}$ at the point $(1, 2)$.

3 A curve is defined by the equation $(x + y)^2 + 2e^y = 11$. Find the value of $\dfrac{dy}{dx}$ at the point $(3, 0)$.

4 Given that $(x^2 - y)^5 = 5\sin y$, find $\dfrac{dy}{dx}$ in terms of x and y.

5 Find the gradient of the curve with implicit equation $x^2 - 2xy + y = 1$ at the point $(2, 1)$.

6 A curve has equation $y^3 + xy - 4x + 2 = 0$.

   (a) Find $\dfrac{dy}{dx}$ in terms of x and y.   /

   (b) Show that the equation of the tangent to the curve at the point $(1, 1)$ is $4y = 3x + 1$.

7 A curve is defined for $x > 0$ by the equation $\sqrt{y} = x^x$.

   (a) Find $\dfrac{dy}{dx}$ in terms of x and y.

   (b) Find the x-coordinate of the stationary point of the curve.

8 A circle whose equation is $x^2 + y^2 + 2x - 6y - 15 = 0$ intersects the negative x-axis at the point A and the positive x-axis at the point B.

   (a) Find the coordinates of A and B.

   (b) Show that $\dfrac{dy}{dx} = \dfrac{x + 1}{3 - y}$.

   (c) Show that the equation of the tangent to the circle at A is $3y + 4x + 20 = 0$.

   (d) The tangents to the circle at the points A and B intersect at the point P. Find the coordinates of P.

9 A circle has equation $x^2 + y^2 - 4x = 6$. Find the coordinates of the two points on the circle where the gradient is 3.

10 (a) Find an equation of the normal to the curve $y^2 = 4x + 1$ at the point $P(2, -3)$.

   (b) This normal intersects the y-axis at the point A and the x-axis at the point B. Find the area of triangle OAB, where O is the origin.

11 A curve has equation $x^2 + y^2 + 3xy - 31 = 0$.

   (a) Find the coordinates of the two points A and B on the curve where $x = 2$.

   (b) Find the equations of the tangents to the curve at the points A and B, giving your answers in the form $ax + by + c = 0$, where a, b and c are integers.

12 The equation of a curve C is $x^2 + y^2 = e^y$.

   (a) Find the coordinates of the two points where the curve crosses the x-axis.

   (b) The normals to C at these two points intersect at the point P. Show that P lies on the y-axis and state its coordinates.

13 A curve with equation $x^2 + x\cos y + 2y - 6 = 0$ intersects the x-axis at the points A and B.

   (a) Find $\dfrac{dy}{dx}$ in terms of x and y.

   (b) The tangents to the curve at A and B intersect at the point R. Find the coordinates of R.

14 The curve C has equation $y^3 - 2y = 4x - x^2$.

   (a) Show that when $\dfrac{dy}{dx} = 0$, $x = 2$.

   (b) Hence, prove that the curve C has only one stationary point and find its coordinates.

15 The equation of a curve is $y = 3e^x y^2 - 4e^{3x}$.

   (a) Find $\dfrac{dy}{dx}$ in terms of x and y.

   (b) The point (a, b) is a stationary point of the curve. Show that $b = 2e^a$.

   (c) Hence find the values of a and b.

Test yourself	What to review
	If your answer is incorrect:
1 Given that $xy = 4x^2 + y^2$, find $\dfrac{dy}{dx}$ giving your answer in terms of x and y.	Review Advancing Maths for AQA C3C4 pages 230–233.
2 (a) Differentiate $\dfrac{x}{x+y}$ with respect to x. **(b)** A curve has implicit equation $y = \dfrac{x}{x+y}$. Find the gradient of the curve at the point $\left(\dfrac{1}{2}, \dfrac{1}{2}\right)$.	Review Advancing Maths for AQA C3C4 pages 230–233.

4

Test yourself (continued)	**What to review**

If your answer is incorrect:

3 The point (p, q) is a stationary point of the curve with equation $2y^2 - x^2y + 8x = 0$.

Review Advancing Maths for AQA C3C4 pages 230–233.

   (a) Use implicit differentiation to find an equation involving $\dfrac{dy}{dx}$, x and y.

   (b) Show that $pq = 4$.

   (c) Hence find the value of p and the value of q.

---

4 Given that $\sin y + \ln(1 + y) - x^2 e^y + x = 0$ is the implicit equation of a curve, find the value of $\dfrac{dy}{dx}$ at the point $(1, 0)$.

Review Advancing Maths for AQA C3C4 pages 230–233.

---

5 A curve is defined by the implicit equation $3x^2y - 3x - 2y^2 = 1$. The line $y = 2$ intersects the curve at the points $(-1, 2)$ and P.

Review Advancing Maths for AQA C3C4 pages 234–239.

   (a) Find the coordinates of P.

   (b) Show that the equation of the tangent to the curve at the point $(-1, 2)$ is $y + 3x + 1 = 0$.

   (c) The tangent at P intersects the tangent $y + 3x + 1 = 0$ at the point Q. Find the coordinates of Q.

---

6 Show that the equation of the normal to the curve $\dfrac{x}{y + 1} = 2x - y - 2$ at the point $(2, 1)$ is $x + 3y = 5$.

Review Advancing Maths for AQA C3C4 pages 234–239.

Test yourself   ANSWERS

5 (a) $\left(\dfrac{3}{2}, 2 \right)$ **(c)** $(1, -4)$

4 1

3 (a) $4y\dfrac{dy}{dx} - 2xy - x^2\dfrac{dy}{dx} + 8 = 0$ **(c)** $p = -2, q = -2$

(b) $\dfrac{1}{3}$

2 (a) $\dfrac{(x + y) - x\left(1 + \dfrac{dy}{dx}\right)}{(x + y)^2}$

1 $\dfrac{dy}{dx} = \dfrac{8x - y}{x - 2y}$

CHAPTER 5

Parametric equations

Key points to remember

1 To find the gradient of a curve defined parametrically in

terms of t, use $\dfrac{dy}{dx} = \dfrac{\frac{dy}{dt}}{\frac{dx}{dt}}$.

2 Parametric equations can be converted to cartesian equations by eliminating the parameter using algebraic rearrangements or trigonometric identities.

Worked example 1

A curve is defined by the parametric equations
$$x = t^2 - 4, \quad y = 1 - t.$$

(a) Find the points on the curve corresponding to
$$t = 0 \text{ and } t = -1.$$

(b) Find the coordinates of the points where the curve crosses the coordinate axes.

(c) Find a cartesian equation for the curve.

(d) Sketch the curve.

(a) When $t = 0$, $x = 0 - 4 = -4$; $y = 1 - 0 = 1$.
This gives the point $(-4, 1)$.

When $t = -1$, $x = 1 - 4 = -3$; $y = 1 + 1 = 2$.
The second point has coordinates $(-3, 2)$.

(b) The curve crosses the x-axis when $y = 0$. This occurs when $t = 1$.

When $t = 1$, $x = 1 - 4 = -3$ so the curve crosses the x-axis at $(-3, 0)$.

The curve crosses the y-axis when $x = 0$. This occurs when $t^2 = 4$.

When $t = 2$, $y = 1 - 2 = -1$ so the curve crosses the y-axis at $(0, -1)$.

When $t = -2$, $y = 1 + 2 = 3$ so the curve also crosses the y-axis at $(0, 3)$.

(c) The equation $y = 1 - t$ can be rearranged to give $t = 1 - y$.

Substituting into $x = t^2 - 4$ gives $x = (1 - y)^2 - 4$.

> You could have left your final answer in this form.

Multiplying out $x = 1 - 2y + y^2 - 4 = y^2 - 2y - 3$.

The curve has cartesian equation $y^2 - 2y = x + 3$.

(d)

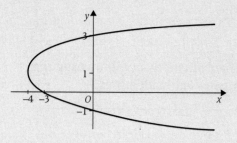

Worked example 2

A curve is defined by the parametric equations
$$x = 4t^3 + 2t + 5, \quad y = 3t^2 - 2t + 7.$$

(a) Find $\dfrac{dy}{dx}$ in terms of t.

(b) Find the equation of the normal to the curve at the point where $t = 1$.

(a) $\dfrac{dx}{dt} = 12t^2 + 2;$

$\dfrac{dy}{dt} = 6t - 2$

Writing $\dfrac{dy}{dx} = \dfrac{\dfrac{dy}{dt}}{\dfrac{dx}{dt}}$

> Using ▇

$\dfrac{dy}{dx} = \dfrac{6t - 2}{12t^2 + 2} = \dfrac{3t - 1}{6t^2 + 1}.$

> You do not need to simplify your answer unless you are asked to prove a given result.

(b) When $t = 1$, the point on the curve is given by
$x = 4 + 2 + 5 = 11$, $\quad y = 3 - 2 + 7 = 8$.

The normal passes through the point $(11, 8)$.

Also $\dfrac{dy}{dx} = \dfrac{3t - 1}{6t^2 + 1} = \dfrac{3 - 1}{6 + 1} = \dfrac{2}{7}$ at this point.

Hence the normal has gradient $-\dfrac{7}{2}$.

The normal has equation $y - 8 = -\dfrac{7}{2}(x - 11)$
or $7x + 2y = 93$.

Worked example 3

A curve is defined by the parametric equations
$$x = 2 - \cot \theta, \quad y = 4 + 3\operatorname{cosec} \theta.$$

(a) Find a cartesian equation for the curve.

(b) Show that $\dfrac{dy}{dx} = -3\cos \theta$.

(a) You can make use of the identity
$\operatorname{cosec}^2 \theta = 1 + \cot^2 \theta$.

Rearranging the parametric equations

$\operatorname{cosec} \theta = \dfrac{y - 4}{3}$ and $\cot \theta = 2 - x$.

Hence $\left(\dfrac{y - 4}{3}\right)^2 = 1 + (2 - x)^2$ is a cartesian equation   • ———— Using **2**

for the curve.

(b) $x = 2 - \cot \theta \Rightarrow \dfrac{dx}{d\theta} = \operatorname{cosec}^2 \theta$

$y = 4 + 3\operatorname{cosec} \theta \Rightarrow \dfrac{dy}{d\theta} = -3\operatorname{cosec} \theta \cot \theta$

Writing $\dfrac{dy}{dx} = \dfrac{\dfrac{dy}{d\theta}}{\dfrac{dx}{d\theta}}$ • ———— Using **1**

$\Rightarrow \dfrac{dy}{dx} = \dfrac{-3\operatorname{cosec} \theta \cot \theta}{\operatorname{cosec}^2 \theta} = \dfrac{-3}{\operatorname{cosec} \theta} \times \dfrac{\cos \theta}{\sin \theta} = -3\cos \theta$.

5

Worked example 4

A curve is defined by the parametric equations
$$x = t^2 + 2, \quad y = t(4 - t^2).$$

(a) **(i)** Find $\dfrac{dy}{dx}$ in terms of t.

(ii) Hence find the gradient of the curve at the point where $t = 2$.

(b) Find a cartesian equation of the curve.

(a) **(i)** $x = t^2 + 2 \Rightarrow \dfrac{dx}{dt} = 2t$

$y = t(4 - t^2) = 4t - t^3 \Rightarrow \dfrac{dy}{dt} = 4 - 3t^2.$

Since $\dfrac{dy}{dx} = \dfrac{\frac{dy}{dt}}{\frac{dx}{dt}}$,   Using **1**

$\dfrac{dy}{dx} = \dfrac{4 - 3t^2}{2t}.$

(ii) When $t = 2$, $\dfrac{dy}{dx} = \dfrac{4 - 12}{4} = -2.$

Therefore, the gradient of the curve at the point where $t = 2$ is equal to -2

(b) Since $x = t^2 + 2$ and $y = t(4 - t^2)$, it is best to square y so that $y^2 = t^2(4 - t^2)^2$.

Rearranging $x = t^2 + 2$ gives $t^2 = x - 2$.

Hence $4 - t^2 = 4 - (x - 2) = 6 - x$.

Therefore $y^2 = (x - 2)(6 - x)^2$ is a cartesian equation of the curve.

Worked example 5

A curve is defined by the parametric equations
$$x = \frac{3}{t} + t^2, \quad y = \frac{1}{t} + 2t^2 \quad (t \neq 0).$$

(a) Verify that $(3y - x)(y - 2x)^2 = 125$ is a cartesian equation for the curve.

(b) Show that $\dfrac{dy}{dx} = \dfrac{4t^3 - 1}{2t^3 - 3}.$

(c) Find an equation for the tangent to the curve at the point where $t = 1$.

(a) $3y - x = \dfrac{3}{t} + 6t^2 - \dfrac{3}{t} - t^2 = 5t^2$

and $y - 2x = \dfrac{1}{t} + 2t^2 - \dfrac{6}{t} - 2t^2 = -\dfrac{5}{t}$

hence $(3y - x)(y - 2x)^2 = 5t^2 \times \dfrac{25}{t^2} = 125$.

Hence a cartesian equation for the curve is

$(3y - x)(y - 2x)^2 = 125$.

(b) $x = \dfrac{3}{t} + t^2 \Rightarrow \dfrac{dx}{dt} = -3t^{-2} + 2t$.

and $y = \dfrac{1}{t} + 2t^2 \Rightarrow \dfrac{dy}{dt} = -t^{-2} + 4t$

Since $\dfrac{dy}{dx} = \dfrac{\frac{dy}{dt}}{\frac{dx}{dt}}$,

| Using **1** |

$\dfrac{dy}{dx} = \dfrac{4t - t^{-2}}{2t - 3t^{-2}}$

$\Rightarrow \dfrac{dy}{dx} = \dfrac{4t^3 - 1}{2t^3 - 3}$

| Multiplying top and bottom by t^2. |

(c) When $t = 1$, $x = 3 + 1 = 4$ and $y = 1 + 2 = 3$.
Tangent passes through the point $(4, 3)$.

Gradient of tangent is equal to the gradient of the curve
when $t = 1$, $\dfrac{dy}{dx} = \dfrac{4 - 1}{2 - 3} = -3$.
Tangent has equation $(y - 3) = -3(x - 4)$ or $3x + y = 15$.

REVISION EXERCISE 5

1 A curve is defined by the parametric equations
$x = t^2 - 9$, $y = 2 + t$.

(a) Find the points on the curve corresponding to $t = 0$ and $t = -1$.

(b) Find the coordinates of the points where the curve crosses the coordinate axes.

(c) Find a cartesian equation for the curve.

(d) Sketch the curve.

2 A curve is defined by the parametric equations
$$x = 5t^3 - 35, \quad y = 2t^5 - 60.$$

  (a) Find $\dfrac{dy}{dx}$ in terms of t.

  (b) Find the gradient of the curve at the point where $t = 2$.

  (c) Find an equation of the normal to the curve at the point where $t = 2$.

3 A curve has parametric equations
$$x = t^2 + 4, \quad y = t^3 - 1.$$

  (a) (i) Find $\dfrac{dx}{dt}$ and $\dfrac{dy}{dt}$.

   (ii) Hence find the gradient of the curve at the point where $t = -1$.

   (iii) Hence find an equation of the tangent to the curve at the point where $t = -1$.

  (b) Find a cartesian equation of the curve.

4 A curve has parametric equations

$$x = 4t + 3, \quad y = 16t(t + 1).$$

  (a) (i) Find $\dfrac{dy}{dx}$ in terms of t.

   (ii) Hence find an equation of the tangent to the curve at the point where $t = 1$.

  (b) Find a cartesian equation of the curve.

5 A curve is defined by the parametric equations
$$x = 4 - t^2, \quad y = t(t^2 + 1).$$

  (a) (i) Find $\dfrac{dy}{dx}$ in terms of t.

   (ii) Hence find an equation of the normal to the curve at the point where $t = -2$.

  (b) Verify that $y^2 = (4 - x)(5 - x)^2$ is a cartesian equation of the curve.

6 A curve is defined by the parametric equations

$$x = 3t - \frac{1}{t}, \quad y = \frac{1}{t} \quad (t \neq 0).$$

(a) Find the coordinates of the point on the curve where $t = \frac{1}{3}$.

(b) Find the gradient of the curve at the point where $t = \frac{1}{3}$.

(c) Find the y-coordinates of the two points on the curve where $x = 2$.

(d) Verify that the cartesian equation of the curve can be written in the form $y^2 + xy = 3$.

7 A curve has parametric equations

$$x = t + \frac{1}{t}, \quad y = t^3 + \frac{1}{t^3} \quad (t \neq 0).$$

(a) (i) Express x^3 in terms of t.

(ii) Hence find a cartesian equation of the curve.

(b) (i) Find $\frac{dx}{dt}$ and $\frac{dy}{dt}$.

(ii) Hence find the gradient of the curve at the point where $t = \sqrt{2}$.

8 A curve is defined by the parametric equations

$$x = 3t^2 - 1, \quad y = 2t(3 - t^2).$$

(a) (i) Find $\frac{dy}{dx}$ in terms of t.

(ii) Hence find an equation of the tangent to the curve at the point where $t = \frac{1}{2}$.

(b) Find a cartesian equation of the curve.

9 A curve is defined by the parametric equations

$$x = 3 + 2\cos\theta, \quad y = 4 - \sin\theta.$$

(a) Find a cartesian equation for the curve.

(b) Find the value of $\frac{dy}{dx}$ at the point where $\theta = \frac{\pi}{4}$.

10 A curve is defined by the parametric equations

$$x = 5 + 2\cos\theta, \quad y = 3\cos 2\theta.$$

(a) Show that $\frac{dy}{dx} = 6\cos\theta$.

(b) Verify that $2(y + 3) = 3(x - 5)^2$ is a cartesian equation for the curve.

5

11 A curve is defined by the parametric equations

$$x = \frac{1}{t-3}, \quad y = \frac{1}{t+1} \quad (t \neq -1, t \neq 3)$$

(a) (i) Express t in terms of x.

(ii) Hence show that the curve has cartesian equation

$$y = \frac{x}{4x+1}.$$

(b) (i) Find $\dfrac{dx}{dt}$ and $\dfrac{dy}{dt}$.

(ii) Hence find the gradient of the curve at the point where $t = 4$.

Test yourself	**What to review**
	If your answer is incorrect:
1 A curve is defined by the parametric equations $x = 1 + t, \quad y = t^2 - 4$. **(a)** Find the points on the curve corresponding to $t = 0$ and $t = 3$. **(b)** Find the coordinates of the points where the curve crosses the coordinate axes. **(c)** Sketch the curve.	Review Advancing Maths for AQA C3C4 pages 240–241.
2 A curve is defined by the parametric equations $x = 3 - 2t + 5t^2, \quad y = t^2 - 4t - 1$. **(a)** Find $\dfrac{dy}{dx}$ in terms of t. **(b)** Find the equation of the tangent to the curve at the point where $t = 1$.	Review Advancing Maths for AQA C3C4 pages 242–244.
3 Find $\dfrac{dy}{dx}$ in terms of θ when **(a)** $x = 5\theta + 2\sin \theta, \quad y = 3 - \cos \theta$ **(b)** $x = 3\cot 3\theta, \quad y = 2\cosec 3\theta$.	Review Advancing Maths for AQA C3C4 pages 245–247.
4 A curve is defined by the parametric equations $x = 2 - t, \quad y = t^3 + 5t$. Find a cartesian equation for the curve.	Review Advancing Maths for AQA C3C4 page 247.

Test yourself *(continued)*	**What to review**

If your answer is incorrect:

5 A curve is defined by the parametric equations

$$x = 3 + \sec\theta, \quad y = 1 - 2\tan\theta.$$

Find a cartesian equation for the curve.

Review Advancing Maths for AQA C3C4 pages 248–249.

---

6 A curve is defined by the parametric equations

$$x = \frac{1}{t} - 2t, \quad y = \frac{2}{t} - t \quad (t \neq 0).$$

(a) Verify that $(x - 2y)(2x - y) = 9$ is a cartesian equation for the curve.

(b) Show that $\dfrac{dy}{dx} = \dfrac{2 + t^2}{1 + 2t^2}$.

(c) Find an equation for the normal to the curve at the point where $t = 1$.

Review Advancing Maths for AQA C3C4 pages 250–255.

5

Test yourself **ANSWERS**

6 (c) $x + y = 0$

5 $(x - 3)^2 = 1 + \dfrac{(1 - y)^2}{4}$

4 $y = (2 - x)^3 + 5(2 - x)$

3 (a) $\dfrac{5 + 2\cos\theta}{\sin\theta}$ **(b)** $\dfrac{2}{3}\cos 3\theta$

2 (a) $\dfrac{5t - 1}{t - 2}$ **(b)** $x + 4y + 10 = 0$

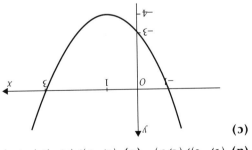

(c)

1 (a) $(1, -4), (4, 5)$ **(b)** $(0, -3), (3, 0), (-1, 0)$

CHAPTER 6

Further trigonometry with integration

Key points to remember

1 $\sin(A \pm B) = \sin A \cos B \pm \cos A \sin B$
$\cos(A \pm B) = \cos A \cos B \mp \sin A \sin B$

These identities are in the formulae booklet.

$$\tan(A \pm B) = \frac{\tan A \pm \tan B}{1 \mp \tan A \tan B}, \quad \{A \pm B \neq (k + \tfrac{1}{2})\pi\}$$

2 $\sin 2A = 2 \sin A \cos A$

$\cos 2A = \cos^2 A - \sin^2 A$, or, using $\cos^2 A + \sin^2 A = 1$,
$\cos 2A = 2 \cos^2 A - 1$
$\cos 2A = 1 - 2 \sin^2 A$

$$\tan 2A = \frac{2 \tan A}{1 - \tan^2 A}$$

3 To integrate either $\sin^2 x$ or $\cos^2 x$ write each in terms of $\cos 2x$.

4 **(a)** $a\cos \theta + b\sin \theta \equiv R\cos(\theta - \alpha)$,

where $R = \sqrt{a^2 + b^2}$ and $\cos \alpha = \dfrac{a}{R}$ and $\sin \alpha = \dfrac{b}{R}$

(b) $a\cos \theta + b\sin \theta \equiv R\sin(\theta + \beta)$,

where $R = \sqrt{a^2 + b^2}$ and $\sin \beta = \dfrac{a}{R}$ and $\cos \beta = \dfrac{b}{R}$

In questions, $R > 0$ and α and β acute will be used.
(Note α and β can be negative in the case when a and b have different signs.)

5 To solve equations of the type $a\cos \theta + b\sin \theta = c$,
as a first step, write $a\cos \theta + b\sin \theta$ either in the form
$R\cos(\theta \pm \alpha)$ or in the form $R\sin(\theta \pm \alpha)$.

Worked example 1

(a) Using the identity $\sin(A \pm B) = \sin A \cos B \pm \cos A \sin B$,

prove that $\sin\left(x + \frac{\pi}{3}\right) + \sin\left(x - \frac{\pi}{3}\right) = \sin x$.

(b) Hence solve the equation $\sin\left(x + \frac{\pi}{3}\right) + \sin\left(x - \frac{\pi}{3}\right) = 0.8$

for $0 \leqslant x \leqslant 2\pi$, giving your answer to two decimal places.

(a) $\sin\left(x + \frac{\pi}{3}\right) + \sin\left(x - \frac{\pi}{3}\right)$ ——————— Using **1** with $A = x$ and $B = \frac{\pi}{3}$

$= \left(\sin x \cos \frac{\pi}{3} + \cos x \sin \frac{\pi}{3}\right) + \left(\sin x \cos \frac{\pi}{3} - \cos x \sin \frac{\pi}{3}\right)$

$= 2\sin x \cos \frac{\pi}{3} = 2\sin x \times \frac{1}{2}$ ——— Calculator set in radian mode gives $\cos \frac{\pi}{3} = 0.5$

$\Rightarrow \sin\left(x + \frac{\pi}{3}\right) + \sin\left(x + \frac{\pi}{3}\right) = \sin x$

(b) $\sin\left(x + \frac{\pi}{3}\right) + \sin\left(x - \frac{\pi}{3}\right) = 0.8 \Rightarrow \sin x = 0.8$ ——— Using part **(a)**

$x = \sin^{-1} 0.8 = 0.927\ldots$ ——— Set calculator in radian mode
In the interval $0 \leqslant x \leqslant 2\pi$, $x = 0.927\ldots,\ \pi - 0.927\ldots$
$\qquad\qquad\qquad\qquad\qquad\qquad x = 0.93^c,\ 2.21^c$ (to 2 d.p.)

6

Worked example 2

(a) Prove the identity $\dfrac{\sin 2x}{1 - \cos 2x} = \cot x,\ \cos 2x \neq 1$.

(b) Hence solve the equation $\dfrac{\sin 2x}{1 - \cos 2x} = 1$ for $0° \leqslant x \leqslant 360°$.

$\cos 2x \neq 1$ is included as the LHS of the identity is not defined when $1 - \cos 2x = 0$

(a) $\dfrac{\sin 2x}{1 - \cos 2x} = \dfrac{2\sin x \cos x}{1 - \cos 2x}$ ——— Using **2**

$= \dfrac{2\sin x \cos x}{1 - (1 - 2\sin^2 x)}$ ——— Since $\cot x = \dfrac{\cos x}{\sin x}$, use $\cos 2x = 1 - 2\sin^2 x$

$= \dfrac{2\sin x \cos x}{2\sin^2 x}$

$\dfrac{\sin 2x}{1 - \cos 2x} = \dfrac{\cos x}{\sin x} = \cot x$ ——— Using **2**

(b) $\dfrac{\sin 2x}{1 - \cos 2x} = 1 \Rightarrow \cot x = 1$ ——— Using identity in part **(a)**

$\Rightarrow \tan x = 1$ ——— Using $\cot x = \dfrac{1}{\tan x}$

$\Rightarrow x = 45°,\ 225°$

Worked example 3

Find $\int (\sin x + 3\cos x)^2\, dx$.

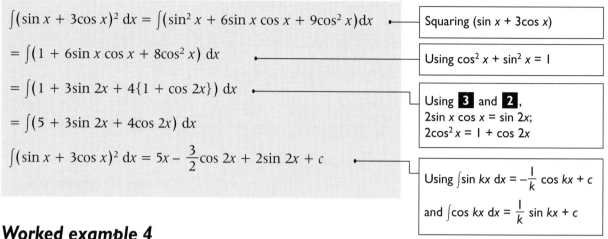

$$\int (\sin x + 3\cos x)^2\, dx = \int (\sin^2 x + 6\sin x \cos x + 9\cos^2 x)\, dx$$

Squaring $(\sin x + 3\cos x)$

$$= \int (1 + 6\sin x \cos x + 8\cos^2 x)\, dx$$

Using $\cos^2 x + \sin^2 x = 1$

$$= \int (1 + 3\sin 2x + 4\{1 + \cos 2x\})\, dx$$

$$= \int (5 + 3\sin 2x + 4\cos 2x)\, dx$$

Using **3** and **2**,
$2\sin x \cos x = \sin 2x$;
$2\cos^2 x = 1 + \cos 2x$

$$\int (\sin x + 3\cos x)^2\, dx = 5x - \frac{3}{2}\cos 2x + 2\sin 2x + c$$

Using $\int \sin kx\, dx = -\frac{1}{k}\cos kx + c$

and $\int \cos kx\, dx = \frac{1}{k}\sin kx + c$

Worked example 4

(a) Express $\cos x + \sqrt{3}\sin x$ in the form $R\cos(x - \alpha)$, where R is a positive constant and α is an acute angle.

(b) State the minimum value of $\cos x + \sqrt{3}\sin x$ and find the values of x in the interval $-360° \leqslant x \leqslant 360°$ at which it occurs.

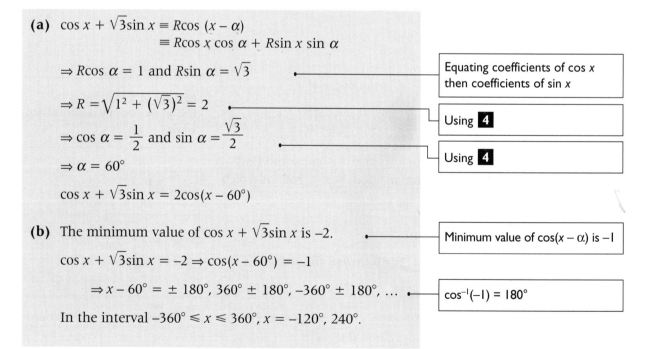

(a) $\cos x + \sqrt{3}\sin x \equiv R\cos(x - \alpha)$
$\equiv R\cos x \cos \alpha + R\sin x \sin \alpha$

$\Rightarrow R\cos \alpha = 1$ and $R\sin \alpha = \sqrt{3}$

Equating coefficients of $\cos x$ then coefficients of $\sin x$

$\Rightarrow R = \sqrt{1^2 + (\sqrt{3})^2} = 2$

Using **4**

$\Rightarrow \cos \alpha = \frac{1}{2}$ and $\sin \alpha = \frac{\sqrt{3}}{2}$

Using **4**

$\Rightarrow \alpha = 60°$

$\cos x + \sqrt{3}\sin x = 2\cos(x - 60°)$

(b) The minimum value of $\cos x + \sqrt{3}\sin x$ is -2.

Minimum value of $\cos(x - \alpha)$ is -1

$\cos x + \sqrt{3}\sin x = -2 \Rightarrow \cos(x - 60°) = -1$

$\Rightarrow x - 60° = \pm 180°, 360° \pm 180°, -360° \pm 180°, \ldots$

$\cos^{-1}(-1) = 180°$

In the interval $-360° \leqslant x \leqslant 360°$, $x = -120°, 240°$.

REVISION EXERCISE 6

1 (a) Write $\sin 20° \cos 30° + \cos 20° \sin 30°$ in the form $\sin p°$.

(b) Write $\cos 40° \cos 25° + \sin 40° \sin 25°$ in the form $\cos q°$.

(c) Write $\dfrac{\tan 34° + \tan 28°}{1 - \tan 34° \tan 28°}$ in the form $\tan r°$.

2 It is given that $y = \sin 3x \cos x - \cos 3x \sin x$.

(a) Simplify $\sin 3x \cos x - \cos 3x \sin x$.

(b) Solve the equation $\sin 3x \cos x = \cos 3x \sin x$ for $-180° < x \leqslant 180°$.

3 A curve has equation $y = (\cos x + \sin x)(\cos x - \sin x)$.

(a) Write $(\cos x + \sin x)(\cos x - \sin x)$ in its simplest form.

(b) Find an equation of the tangent to the curve at the point P on the curve where $x = \dfrac{\pi}{4}$.

(c) The tangent to the curve at P intersects the y-axis at the point A. Show that the area of triangle OAP, where O is the origin, is $\dfrac{\pi^2}{16}$.

4 (a) Find $\int 2 \sin^2 x \, dx$.

(b) Hence evaluate $\displaystyle\int_0^{\frac{\pi}{12}} 2 \sin^2 x \, dx$

5 (a) Express $\sqrt{3}\sin x + \cos x$ in the form $R\sin(x + \alpha)$, where R is a positive constant and α is an acute angle.

(b) State the maximum value of $\sqrt{3}\sin x + \cos x$ and find the values of x in the interval $-360° \leqslant x \leqslant 360°$ at which it occurs.

6 It is given that $5\sin 2\theta = 8\sin\theta$.

(a) Use the identity $\sin 2\theta = 2\sin\theta \cos\theta$ to show that $\sin\theta = 0$ or $\cos\theta = \dfrac{4}{5}$

(b) Hence, or otherwise, solve the equation $5\sin 2\theta = 8\sin\theta$ giving all values of θ to the nearest $0.1°$ in the interval $0° \leqslant \theta \leqslant 360°$.

6

7 Solve the equation $2\sin 2\theta = \cos\theta$ giving all values of θ to the nearest degree in the interval $0° \leqslant \theta \leqslant 360°$.

8 (a) Express $\cos 2x$ in terms of $\cos x$ and $\sin x$.

(b) Hence solve the equation $\cos^2 x = 1 + \sin^2 x$
for $0° \leqslant x \leqslant 360°$.

9 (a) Express $12\cos x - 5\sin x$ in the form
$R\cos(x + \alpha)$, where R is a positive constant and α is an acute angle. Give your value of α to the nearest $0.1°$.

(b) Solve the equation $12\cos x - 5\sin x = 7.8$ in the interval $-180° \leqslant x \leqslant 180°$.

10 A curve is defined for $0 \leqslant x \leqslant 2\pi$ by the equation
$y = 6\sin x - 8\cos x$

(a) Express $6\sin x - 8\cos x$ in the form $R\sin(x - \alpha)$,

where R is a positive constant and $0 < \alpha < \dfrac{\pi}{2}$. Give your value of α to three significant figures.

(b) (i) Hence find $\dfrac{dy}{dx}$.

(ii) Find the x-coordinates of the points on the curve where the tangent to the curve is parallel to the line $y = 4x + 3$. Give your answers to three significant figures.

11 (a) Show that $(\cos x + \sin x)^2 = 1 + \sin 2x$.

(b) Hence solve the equation $(\cos x + \sin x)^2 = 0.5$ for $-180° \leqslant x \leqslant 180°$.

(c) Show that $\displaystyle\int_0^{\frac{\pi}{2}} (\cos x + \sin x)^2 \, dx = \dfrac{\pi + 2}{2}$.

12 Find $\displaystyle\int (\tan x + 2\cos x)^2 \, dx$

13 (a) By writing $3x$ as $(2x+x)$, show that
$\cos 3x = \cos x \, (1 - 4\sin^2 x)$.

(b) (i) Hence find $\displaystyle\int \dfrac{\cos 3x}{\cos x} \, dx$.

(ii) Show that $\displaystyle\int_0^{\frac{\pi}{4}} \dfrac{\cos 3x}{\cos x} \, dx = 1 - \dfrac{\pi}{4}$.

14 (a) (i) By expanding $(\sin^2 \theta + \cos^2 \theta)^2$, prove the identity

$$\sin^4 \theta + \cos^4 \theta = 1 - \frac{1}{2}\sin^2 2\theta.$$

(ii) Hence prove that $4(\sin^4 \theta + \cos^4 \theta) = \cos 4\theta + 3$.

(b) Solve the equation $\sin^4 \theta + \cos^4 \theta = \frac{7}{8}$ for $0° < \theta < 180°$

15 (a) Express $\tan 2x$ in terms of $\tan x$.

(b) Use the identity $\tan(A + B) = \dfrac{\tan A + \tan B}{1 - \tan A \tan B}$,

with $A = 2x$ and $B = x$ to show that

$$\tan 3x = \frac{3\tan x - \tan^3 x}{1 - 3\tan^2 x}.$$

(c) Hence solve the equation $\tan 3x = 4\tan x$ giving values of x in radians in the interval $0 \le x \le 2\pi$.

16 The region bounded by the curve $y = 1 + 2\sin x$, the x-axis and the lines $x = 0$ and $x = \dfrac{\pi}{3}$ is rotated through 2π radians about the x-axis. Given that $\sin \dfrac{2\pi}{3} = \dfrac{\sqrt{3}}{2}$, show that the volume of the solid formed is $\pi\left(\pi + 2 - \dfrac{\sqrt{3}}{2}\right)$.

6

Test yourself	What to review
	If your answer is incorrect:
1 Prove the identity $\dfrac{2\cos(A + B)}{\sin(A + B) + \sin(A - B)} = \cot A - \tan B$.	Review Advancing Maths for AQA C3C4 pages 256–259.
2 Solve the equation $\dfrac{\tan 4x - \tan x}{1 + \tan 4x \tan x} = 1$ for $0° \le x \le 180°$.	Review Advancing Maths for AQA C3C4 pages 256–259.
3 (a) Show that $\cos 2x - 3\cos x - 4$ can be written in the form $(2\cos x - 5)(\cos x + 1)$. **(b)** Hence solve the equation $\cos 2x = 3\cos x + 4$ for $0° \le x \le 360°$.	Review Advancing Maths for AQA C3C4 pages 260–262.

Test yourself (continued)

What to review

If your answer is incorrect:

4 (a) Show that the substitution $x = 2\sin\theta$ transforms $\int\sqrt{4 - x^2}\ dx$ to $\int 4\cos^2\theta\ d\theta$.

Review Advancing Maths for AQA C3C4 pages 263–264.

(b) Hence find the exact value of $\int_0^2\sqrt{4 - x^2}\ dx$.

5 (a) Express $\cos x - 2\sqrt{2}\sin x$ in the form $R\cos(x + \alpha)$, where R is a positive constant and $0 < \alpha < \dfrac{\pi}{2}$.

Give your value of α to three significant figures.

Review Advancing Maths for AQA C3C4 pages 265–268.

(b) State the maximum value of $\cos x - 2\sqrt{2}\sin x$.

(c) Describe a sequence of geometrical transformations that maps the graph of $y = \cos x$ onto the graph of $y = \cos x - 2\sqrt{2}\sin x$.

6 (a) Express $\sin x + \cos x$ in the form $R\sin(x + \alpha)$, where R is a positive constant and α is an acute angle.

Review Advancing Maths for AQA C3C4 pages 269–271.

(b) Hence solve the equation $\sqrt{2}(\sin x + \cos x) = 1$ for $-180° \leqslant x \leqslant 180°$.

Test yourself ANSWERS

6 (a) $\sqrt{2}\sin(x + 45°)$ **(b)** $-15°, 105°$

(c) Translation $\begin{bmatrix} -\alpha \\ 0 \end{bmatrix}$, stretch in y-direction scale factor 3.

5 (a) $3\cos(x + 1.23)$ **(b)** 3

4 (b) π

3 (b) 180°

2 15°, 75°, 135°

CHAPTER 7

Exponential growth and decay

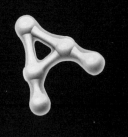

Key points to remember

1 The solution of the equation is $a^x = b$ is $x = \dfrac{\ln b}{\ln a}$, when logarithms are taken to base e.

2 The solution of the equation $a^x = b$ is $x = \dfrac{\log_{10} b}{\log_{10} a}$, when logarithms are taken to base 10.

3 The formula $x = a \times b^{kt}$, where a, b and k are positive constants, indicates that x is growing exponentially.

4 The formula $x = a \times b^{-kt}$, where a, b and k are positive constants, indicates that x is decaying exponentially.

5 In general, $x = Ae^{kt} \Rightarrow \dfrac{dx}{dt} = kx$.

6 The general solution of $\dfrac{dx}{dt} = kx$ is $x = Ae^{kt}$, where A is an arbitrary constant.

7

Worked example 1

Solve the equation $5^{2x-1} = 19$ giving your answer to four significant figures.

> This type of question has already been tested in C2 but is included here because you need to use this technique in questions on growth and decay.

$5^{2x-1} = 19$

$\ln(5^{2x-1}) = \ln 19$ •————————

> Taking natural logarithms of both sides.

$\Rightarrow (2x-1)\ln 5 = \ln 19$

$\Rightarrow 2x - 1 = \dfrac{\ln 19}{\ln 5}$ •————————

> Or using **1**

$\Rightarrow 2x - 1 = 1.82948\ldots$ •————————

> Make sure you do NOT evaluate $\ln\left(\dfrac{19}{5}\right)$.

$\Rightarrow 2x = 2.829\ldots$

$\Rightarrow x = 1.415$ (to four significant figures)

Worked example 2

The mass, m grams, of a melting snowball at time t minutes after it is placed on a wall, is given by $m = 550e^{-kt}$, where k is a constant.

(a) State the initial mass of the snowball.

(b) Given that the snowball has mass 460 grams after 25 minutes, find the value of k.

(c) Find the time taken, to the nearest minute, for the mass of the snowball to reduce to 300 grams.

(a) Initial mass is given by putting $t = 0$.
$m = 550e^0 = 550$.
Initial mass is 550 grams.

> Remember that $e^0 = 1$

(b) $m = 460$ when $t = 25$.
Substituting into $m = 550e^{-kt}$ gives $460 = 550e^{-25k}$

$\Rightarrow e^{25k} = \dfrac{550}{460}$

$\Rightarrow 25k = \ln\left(\dfrac{550}{460}\right) \quad \Rightarrow k = \dfrac{1}{25}\ln\left(\dfrac{55}{46}\right) \approx 0.00715$.

(c) Substituting $m = 300$ into $m = 550e^{-kt}$ gives

$300 = 550e^{-kt} \quad \Rightarrow e^{kt} = \dfrac{550}{300}$

$\Rightarrow kt = \ln\left(\dfrac{55}{30}\right) \quad \Rightarrow t = \dfrac{1}{k}\ln\left(\dfrac{55}{30}\right) \approx 84.8$

Time taken is 85 minutes, to nearest minute.

Worked example 3

(a) Given that $P = Ae^{kt}$, where A and k are constants, show that
$$\frac{dP}{dt} = kP.$$

(b) The rate of increase of a population, P, of insects in a colony at time t is given by $\dfrac{dP}{dt} = kP$, where k is a positive constant.

    (i) When $t = 0$, $P = 150$ and when $t = 2$, $P = 230$. Find the value of k.

    (ii) Hence determine the population of insects when $t = 5$.

(a) $P = Ae^{kt} \Rightarrow \dfrac{dP}{dt} = Ake^{kt} = k(Ae^{kt})$

$\Rightarrow \dfrac{dP}{dt} = kP$

(b) Using the result from part **(a)**, the general solution of

$\dfrac{dP}{dt} = kP$ is $P = Ae^{kt}$.

> Using **6**

   (i) When $t = 0$, $P = 150 \Rightarrow 150 = Ae^0 \Rightarrow A = 150$.
       When $t = 2$, $P = 230 \Rightarrow 230 = 150e^{2k}$

$$\Rightarrow 2k = \ln\left(\frac{230}{150}\right) \Rightarrow k = \frac{1}{2}\ln\left(\frac{23}{15}\right) \approx 0.2137.$$

   (ii) $P = 150e^{kt}$. So when $t = 5$, $P = 436.6995\ldots$

The population could be interpreted as 436 (since that is the largest integer less than the value of P)
or 437 (giving your answer to the nearest integer).

This is why an examination question is likely to ask for the population to be given to two significant figures so that your answer could then be given as 440.

> This quite often happens in a modelling situation where you have to interpret your final answer as an integer.

7

Worked example 4

The amount of money, £Q, in a special bank account grows at a rate proportional to the amount in that account at any moment in time. Initially £1 000 is deposited and after 7 years it grows to £1 620.

(a) Find an expression for Q in terms of t, the time in years of the investment.

(b) Determine how long it takes to have £2 500 in the account. Give your answer to the nearest month.

(a) The statement at the beginning of the question can be interpreted mathematically as $\dfrac{dQ}{dt} = kQ$, where k is a constant.

Hence $\dfrac{dt}{dQ} = \dfrac{1}{kQ}$ and integrating with respect to Q gives

$t = \dfrac{1}{k}\ln Q + C$, where C is a constant.

> You will learn how to solve this type of differential equations by separating the variables in the next chapter.

> This approach uses the fact that
> $$\frac{dy}{dx} = \frac{1}{\frac{dx}{dy}}$$

When $t = 0$, $Q = 1000$, hence $C = -\dfrac{1}{k}\ln 1000$

So the solution to the differential equation is

$$t = \frac{1}{k}\ln Q - \frac{1}{k}\ln 1000 = \frac{1}{k}\ln\left(\frac{Q}{1000}\right)$$

Hence $kt = \ln\left(\dfrac{Q}{1000}\right)$ or $e^{kt} = \left(\dfrac{Q}{1000}\right)$

Therefore $Q = 1000e^{kt}$

When $t = 7$, $Q = 1620 \Rightarrow 1620 = 1000e^{7k} \Rightarrow$

$1.62 = e^{7k} \Rightarrow 7k = \ln 1.62 \Rightarrow k = \dfrac{1}{7}\ln 1.62 \approx 0.0689$

The required expression is $Q = 1000e^{kt}$,

where $k = \dfrac{1}{7}\ln 1.62 \approx 0.0689$.

(b) From a previous line in the solution to part **(a)**,

$$kt = \ln\left(\frac{Q}{1000}\right).$$

When $Q = 2\,500$, $kt = \ln\left(\dfrac{2\,500}{1\,000}\right) \approx 0.9163$

> Or by taking natural logarithms of both sides of the solution above.

Therefore $t \approx 13.2988$

Assuming interest is added at least monthly, time taken is 13 years and 4 months to have £2 500 in the account.

> Note that after 13 yrs 3 months, when $t = 13.25$ there will not yet be £2 500 in the account.

REVISION EXERCISE 7

1 Solve the equations **(a)** $3^x = 58$ **(b)** $19^x = 4$ **(c)** $7^x = 0.37$, giving your answers to three decimal places.

2 Given that $2^{4x+3} = 27$, show that $x = \dfrac{1}{4}\left(\dfrac{3\ln 3}{\ln 2} - 3\right)$.

3 The amount of money, £P, in a bank account at time t months after the account is opened is given by $P = 200 \times 1.003^t$.

(a) Find the initial amount of money in the account.

(b) Find the amount of money, to the nearest penny, in the account after one year.

(c) Determine the length of time, to the nearest month, when there will first be £500 in the account.

4 The temperature $T°C$ of liquid in a container m minutes after a particular instant is given by $T = 25 + 50 \times 3^{-0.04m}$.

(a) State the initial temperature of the liquid.

(b) Calculate its temperature after 10 minutes, giving your answer to three significant figures.

(c) Find the time taken for the temperature to fall to $35°C$, giving your answer in minutes to three significant figures.

5 (a) Given that $P = Ae^{5t}$, where A is a constant, show that
$$\frac{dP}{dt} = 5P.$$

(b) The rate of increase of a population, P, at time t is given by $\frac{dP}{dt} = 5P$.

> Hence, you may assume that $P = Ae^{5t}$, where A is a constant.

Given that the population was 3000 when $t = 0$, find the value of P when $t = 2$, giving your answer to the nearest million.

6 (a) (i) Given that $\frac{dN}{dt} = -\frac{N}{10}$, show that $-0.1\frac{dt}{dN} = \frac{1}{N}$.

(ii) Hence, by integrating, show that $N = Ae^{-0.1t}$, where A is a constant.

(b) The number of particles at time t of a certain radioactive substance is N.

The substance decays in such a way that $\frac{dN}{dt} = -\frac{N}{10}$.

Given that when $t = 0$, N was equal to 2000, find the value of t when N is reduced to 500, giving your answer to three significant figures.

7 (a) Given that $P = Ae^{kt}$, where A and k are constants, show that $\frac{dP}{dt} = kP$.

(b) The rate of increase of a population, P, of micro-organisms at time t is given by $\frac{dP}{dt} = kP$, where k is a positive constant.

Hence, you may assume that $P = Ae^{kt}$, where A and k are constants.

Given that when $t = 0$, $P = 200$ and when $t = 1$, $P = 630$, find the value of P when $t = 3$, giving your answer to three significant figures.

8 (a) Prove that the general solution of $\frac{dN}{dt} = kN$, is $N = Ae^{kt}$, where A and k are constants.

7

You may wish to solve the differential equations in the remaining questions by separating the variables (discussed in the next chapter) instead of using the technique shown in worked example **4(a)**. You need to be able to obtain the general solution to problems involving growth and decay using integration rather than simply writing down the general solution.

(b) The number of bacteria, N, in a culture increases at a rate proportional to N. At noon there are 400 bacteria, and 30 minutes later there are 625 bacteria. Find the time, to the nearest minute, at which the number of bacteria will reach 1000.

9 A radioactive substance decays at a rate proportional to its mass m. Initially its mass was 7×10^{-3} g and after 40 days its mass reduced to 6.3×10^{-3} g.

(a) Find an expression for m in terms of the time t, in days from when the mass was 7×10^{-3} g.

(b) How long would it take for the mass to reach two thirds its original size?

10 An experiment reveals that the number, N, of bacteria in a test tube increases at a rate proportional to the number present at any instant. Initially there are 700 present and after 20 days the number has risen to 3 000.

(a) Find an expression for N in terms of the time, t, that has elapsed in days since the experiment began.

(b) Find the number of days required for the number of bacteria to increase to 7 000.

11 The yearly rate of increase of the population, P, in a certain country is proportional to P. On January 1st 2000 there were 12 million and the population rose to 14 million six years later.

(a) Find an expression for P in terms of the time, t, measured in years from January 1st 2000.

(b) Predict the population of the country on January 1st 2015, giving your answer to three significant figures.

12 The amount of money, £ M, in a particular bank account grows at a rate proportional to the amount in that account at any moment in time. Initially £500 is deposited and after 8 years it grows to £620.

   (a) Find an expression for M in terms of t, the time in years of the investment.

   (b) Determine how long it takes to have £1000 in the account.

13 The rate at which a particular car depreciates is proportional to the value of the car at that instant. The car cost £15000 when new and was worth £11000 after two years.

   (a) Predict the value, to the nearest £1000, of the car after a further two years.

   (b) Calculate the length of time from new, to the nearest year, for the car only to be worth £5000.

Test yourself	What to review
	If your answer is incorrect:
1 Solve the equation $7^x = 53$, giving your answer to three significant figures.	Review Advancing Maths for AQA C3C4 pages 275–276.
2 Given that $3^{5x+7} = 49$, show that $x = \dfrac{1}{5}\left(\dfrac{2\ln 7}{\ln 3} - 7\right)$.	Review Advancing Maths for AQA C3C4 pages 276–278.
3 The population, P million, for a particular country is given by $P = 19.5 \times 1.03^t$, where t is the number of years after January 1st 2006. Use the formula to predict: **(a)** the population on January 1st 2009, **(b)** the year when the population will first reach 30 million.	Review Advancing Maths for AQA C3C4 pages 278–280.
4 Given that $N = 5000e^{-\frac{t}{10}}$, **(a)** show that $\dfrac{dN}{dt} = kN$ and state the value of k, **(b)** use the formula to find the value of t when $N = 3500$, giving your answer to three significant figures.	Review Advancing Maths for AQA C3C4 pages 281–283.

7

Test yourself *(continued)*	What to review
	If your answer is incorrect:

5 **(a)** Show that the general solution of $\dfrac{dN}{dt} = kN$ is

$N = Ae^{kt}$, where A and k are constants.

Review Advancing Maths for AQA C3C4 pages 281–283.

(b) The number of bacteria, N, in a colony is such that the rate of increase of N is proportional to N.
The time, t, is measured in hours from the instant that $N = 5$ million.
When $t = 3$, $N = 7$ million. Find the value of t when $N = 8$ million, giving your answer to three significant figures.

Test yourself ANSWERS

5 **(b)** 4.19

4 **(a)** $k = -\dfrac{1}{10}$ **(b)** 3.57

3 **(a)** 21.3 million   **(b)** 2020

1 2.04

CHAPTER 8
Differential Equations

Key points to remember

1 A **differential equation** is an equation which involves at least one derivative of a variable with respect to another variable.

For example $\quad \dfrac{dy}{dx} = 2x - 3, \quad x\dfrac{dx}{dt} = e^t \sin x, \quad \dfrac{d^2m}{dt^2} + 3\dfrac{dm}{dt} + 4m = e^t$

2 **First order differential equations** are differential equations in which the highest derivative is the first.

For example $\dfrac{dy}{dx} = 2x^4 + 3, \quad x^2\dfrac{dx}{dt} = \ln t$

3 • The rate of **increase** of x is proportional to $x \Rightarrow \dfrac{dx}{dt} = kx$

   • The rate of **decrease** of x is proportional to $x \Rightarrow \dfrac{dx}{dt} = -kx$
   where k is a constant of proportionality.

4 The **general solution** of the first order differential equation $\dfrac{dy}{dx} = \text{g}(x)\text{h}(y)$ is given by

$$\int \frac{1}{\text{h}(y)}\, dy = \int \text{g}(x)dx + A, \text{ where } A \text{ is an arbitrary constant.}$$

5 To find a solution of a differential equation which satisfies a given condition, firstly find the general solution of the differential equation and then substitute the given condition to find the arbitrary constant.

8

Worked example 1

The gradient of a curve at each point (x, y) on the curve is equal to the product of the squares of the coordinates of that point.

(a) Express this information in the form of a differential equation.

(b) Given that the curve passes through the point (3, 1) find the equation of the curve, giving your answer in the form $y = \text{f}(x)$.

(c) The curve intersects the y-axis at the point P. Find the coordinates of P.

(a) $\dfrac{dy}{dx} = x^2 y^2$

> The gradient of the curve at (x, y) is $\dfrac{dy}{dx}$

(b) $\Rightarrow \dfrac{1}{y^2} \dfrac{dy}{dx} = x^2$

$\Rightarrow \displaystyle\int \dfrac{1}{y^2}\, dy = \int x^2\, dx$

> Using **4** to separate the variables.

$\Rightarrow -\dfrac{1}{y} = \dfrac{1}{3}x^3 + A$

> The general solution

When $x = 3$, $y = 1 \Rightarrow -1 = \dfrac{27}{3} + A \Rightarrow A = -10$

> Using **5** since $(3, 1)$ lies on the curve.

The equation of the curve is $-\dfrac{1}{y} = \dfrac{1}{3}x^3 - 10$

> The question states the form $y = f(x)$, in which the equation of the curve is to be given. If no particular form had been stated, this answer would be acceptable.

$\Rightarrow -\dfrac{1}{y} = \dfrac{x^3 - 30}{3} \Rightarrow -\dfrac{y}{1} = \dfrac{3}{x^3 - 30}$

> $\dfrac{a}{b} = \dfrac{c}{d} \Rightarrow \dfrac{b}{a} = \dfrac{d}{c}$

$\Rightarrow y = -\dfrac{3}{x^3 - 30}$

> Write the equation of the curve in the form $y = f(x)$.

(c) At P, $x = 0 \Rightarrow y = -\dfrac{3}{0 - 30} = \dfrac{1}{10}$

> Put $x = 0$ in the equation of the curve.

P is the point $\left(0, \dfrac{1}{10}\right)$.

Worked example 2

The number of bacteria, x, in a colony is such that the rate of increase of x is proportional to x. The time, t, is measured in hours from the instant that $x = 2000$.

(a) By forming and solving a differential equation, show that $x = 2000e^{kt}$, explaining why the constant k is positive.

(b) Find an expression for t in terms of k at which the rate of increase of x is twice its initial value.

(c) Given that when $t = 3$, $x = 4800$ find the value of t when $x = 6000$, giving your answer to three significant figures.

(a) 'Rate of **increase** of x is proportional to x' $\Rightarrow \dfrac{dx}{dt} = kx$,

> Using **3**

Since x is **increasing** $\dfrac{dx}{dt}$ is positive; also the number of bacteria x is always >0 so k, the constant of proportionality must be positive.

$$\frac{dx}{dt} = kx \Rightarrow \frac{1}{x}\frac{dx}{dt} = k \Rightarrow \int \frac{1}{x}\,dx = \int k\,dt$$

> Using **4** to separate the variables

$$\Rightarrow \ln x = kt + c$$

$$\Rightarrow x = e^{kt+c}$$

> Using $\ln x = N \Rightarrow x = e^{N}$

$$\Rightarrow x = e^{kt} \times e^{c}$$

So the general solution is $x = Ae^{kt}$

> Writing A for the constant e^{c}

When $t = 0$, $x = 2000$

> Since 't is measured from the instant that $x = 2000$'

$$\Rightarrow 2000 = A$$

So $\qquad\qquad x = 2000e^{kt}$

> Since $e^{0} = 1$

(b) Rate of increase of x is $\dfrac{dx}{dt} = 2000k\,e^{kt}$.

> $\dfrac{d}{dt}(e^{kt}) = ke^{kt}$

Initially $\dfrac{dx}{dt} = 2000k$

> Using $\dfrac{dx}{dt} = kx$ with initial value of x

When $\dfrac{dx}{dt} = 4000k$, $2000k\,e^{kt} = 4000k$

> Need to find t when $\dfrac{dx}{dt} = 2 \times 2000k$

$$\Rightarrow e^{kt} = 2$$

$$\Rightarrow kt = \ln 2$$

> $e^{N} = p \Rightarrow N = \ln p$

$$\Rightarrow t = \frac{\ln 2}{k}$$

(c) $\qquad\qquad\qquad\qquad x = 2000e^{kt}$

When $t = 3$, $x = 4800$ $\Rightarrow 4800 = 2000e^{3k}$

$$\Rightarrow e^{3k} = 2.4 \quad \Rightarrow 3k = \ln 2.4$$

> At this stage, leave k in an exact form.

When $x = 6000$ $\qquad \Rightarrow 6000 = 2000e^{kt}$

$$\Rightarrow e^{kt} = 3 \qquad \Rightarrow kt = \ln 3$$

$$\Rightarrow t = \frac{3\ln 3}{\ln 2.4}$$

> Eliminating k, $\dfrac{kt}{3k} = \dfrac{\ln 3}{\ln 2.4}$

$$\Rightarrow t = \frac{3.2958\ldots}{0.8754\ldots} = 3.764\ldots$$

$$\Rightarrow t = 3.76 \text{ (to 3sf)}$$

8

Worked example 3

Find the general solution of the differential equation $\frac{dy}{dx} = y \tan x$, giving your answer in the form $y = f(x)$.

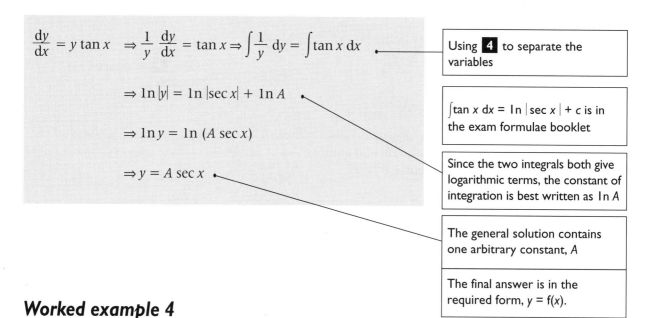

$$\frac{dy}{dx} = y \tan x \implies \frac{1}{y}\frac{dy}{dx} = \tan x \implies \int \frac{1}{y}\,dy = \int \tan x\,dx$$

Using **4** to separate the variables

$$\implies \ln|y| = \ln|\sec x| + \ln A$$

$\int \tan x\,dx = \ln|\sec x| + c$ is in the exam formulae booklet

$$\implies \ln y = \ln(A \sec x)$$

Since the two integrals both give logarithmic terms, the constant of integration is best written as $\ln A$

$$\implies y = A \sec x$$

The general solution contains one arbitrary constant, A

The final answer is in the required form, $y = f(x)$.

Worked example 4

Given that $\frac{dy}{dx} = 6x^2 \sqrt{(y+1)^3}$ and when $x = 1, y = 0$, find the value of y when $x = 0$.

$$\frac{dy}{dx} = 6x^2 (y+1)^{\frac{3}{2}}$$

$$\implies \frac{1}{(y+1)^{\frac{3}{2}}}\frac{dy}{dx} = 6x^2 \qquad \implies \int (y+1)^{-\frac{3}{2}}\,dy = \int 6x^2\,dx$$

Using **4** to separate the variables

$$\implies -2(y+1)^{-\frac{1}{2}} = 2x^3 + c$$

When $x = 1, y = 0$
$$\implies -2 = 2 + c \implies c = -4$$

$$\implies -(y+1)^{-\frac{1}{2}} = x^3 - 2$$

$$\frac{1}{\sqrt{y+1}} = 2 - x^3$$

This is the solution of the differential equation that satisfies the boundary condition $x = 1, y = 0$.

When $x = 0$, $\quad \sqrt{y+1} = \frac{1}{2} \implies y + 1 = \frac{1}{4} \implies y = -\frac{3}{4}$

REVISION EXERCISE 8

1 Find the general solution of the differential equation

$$\frac{dy}{dx} = \frac{x + 4}{y}.$$

2 The gradient of a curve at the point (x, y) is given by the differential equation $\frac{dy}{dx} = \frac{2 - x}{y}$.

  (a) Find the general solution of $\frac{dy}{dx} = \frac{2 - x}{y}$.

  (b) Given that the curve passes through the point $(4, 2)$ show that the curve is a circle and find its radius and the coordinates of its centre.

3 (a) Given that $\frac{4}{x(x + 2)} \equiv \frac{C}{x} + \frac{D}{x + 2}$ find the values of C and D.

  (b) (i) Find the general solution of the differential equation $\frac{dy}{dx} = \frac{4y}{x(x + 2)}$, giving your answer in the form $y = f(x)$.

  (ii) Given that $y = 2$ when $x = 1$ find the value of y when $x = 10$.

4 An antique is valued at £10 000 on January 1st 2000. A mathematical model suggests that t years later its value, £V, increases at a rate proportional to the square root of its value.

  (a) Write down a differential equation in $\frac{dV}{dt}$ to represent this model.

  (b) Show that $2\sqrt{V} = kt + 200$, where k is a positive constant.

  (c) On 1st January 2005 the antique is valued at £14 400. Find the year in which the value of the antique will first exceed £22 500.

5 (a) Given that $y = -1$ when $t = 0$, solve the differential equation $\frac{dy}{dt} = \frac{8 - y}{4}$.

  (b) Find the value of t when $y = 3$ giving your answer to three significant figures.

  (c) Find an expression for y in terms of t.

  (d) State the limiting value of y as $t \to \infty$.

8

6 (a) Use integration by parts to find $\int t \cos t \, dt$.

(b) Given that $\dfrac{dx}{dt} = t \cos t \sqrt{x}$ and $x = 1$ when $t = 0$,

find x when $t = \dfrac{\pi}{2}$, leaving your answer in terms of π.

7 (a) Use the substitution $u = 1 + \tan x$, or otherwise, to

determine $\displaystyle\int \frac{1}{(1 + \tan x)^2 \cos^2 x} \, dx$.

(b) (i) Hence find the solution of the differential equation

$$(1 + \tan x)^2 \cos^2 x \, \frac{dy}{dx} = \sqrt{y}, \text{ given that}$$

$y = 1$ when $x = 0$.

(ii) Find the value of y when $x = \dfrac{\pi}{4}$.

8 (a) Water is flowing into a container at a constant rate of R cm³ per second.

Express this information in the form of a differential equation.

(You should define any variables used.)

(b) Water is flowing into a container at a constant rate of 24 cm³ per second and at the same time water is leaking from the container at a rate proportional to the volume of the water in the container. The container is initially empty and t seconds later the volume of water in the container is V cm³. This situation can be modelled by

the differential equation $\dfrac{dV}{dt} = c - kV$, where c and k are constants.

(i) State the value of c.

(ii) Solve the differential equation to show that

$$V = \frac{24}{k}\left(1 - e^{-kt}\right).$$

(iii) Given also that the rate of increase of water in the container is 12 cm³ per second when $t = 4$, find the volume of water in the container when $t = 12$.

9 At each point (x, y) on a curve C, the gradient of the curve is given by $\dfrac{dy}{dx} = \dfrac{1 + x}{y}$.

The point $A(-1, 2)$ lies on C.

(a) Show that $\dfrac{d^2y}{dx^2} = \dfrac{y^2 - (1 + x)^2}{y^3}$.

(b) Verify that A is a stationary point and determine whether it is a maximum or a minimum.

(c) Find the equation of the curve C giving your answer in the form $y^2 = f(x)$.

10 (a) (i) Write $\dfrac{2}{(x + 1)(x + 3)}$ in the form $\dfrac{A}{x + 1} + \dfrac{B}{x + 3}$.

    (ii) Find the general solution of the differential

        equation $2\dfrac{dx}{dt} = (x + 1)(x + 3)$.

(b) Given that $x = 1$ when $t = 0$, show that $x = \dfrac{3e^t - 2}{2 - e^t}$.

(c) Find the value of x when $t = \ln 4$.

(d) Find the value of t when $x = 7$, giving your answer in the form $\ln p$.

11 An oil slick in the sea is approximately circular and when first observed its radius is 50 metres and the radius is increasing at a rate of 2 metres per minute. At time t minutes after being first observed its radius is r metres.

It is believed that the radius will increase at a rate

proportional to $\dfrac{1}{r^2(1 + t)}$.

(a) Show that the differential equation for this situation is

$$\frac{dr}{dt} = \frac{5000}{r^2(1 + t)}.$$

(b) Solve the differential equation.

(c) Use your solution to predict the radius of the oil slick 3 hours after being first observed, giving your answer to two significant figures.

8

Test yourself	**What to review**
	If your answer is incorrect:
1 (a) The gradient of a curve at each point (x, y) on the curve is equal to the square root of the product of the coordinates of that point. Express this information in the form of a differential equation.	Review Advancing Maths for AQA C3C4 pages 291–294.
(b) Water flows into a container at a constant rate of P cm^3 per second and leaks out at a rate which is proportional to the square of the volume, V cm^3, at time t seconds. Write down an expression for $\dfrac{dV}{dt}$.	
2 Find the general solution of the differential equation $\dfrac{dx}{dt} = 2\cos^2 x \cos^2 t$.	Review Advancing Maths for AQA C3C4 pages 294–297.

Test yourself *(continued)*	What to review

If your answer is incorrect:

3 Find the solution of the differential equation

$$x\frac{dy}{dx} = x^2 y + y \text{ given that } y = 4 \text{ when } x = 2.$$

Give your answer in the form $y = f(x)$.

Review Advancing Maths for AQA C3C4 pages 297–300.

---

4 In a model to estimate the decrease in the mass of a snowball it is assumed that its mass m at time t minutes decreases at a rate which is proportional to m.

(a) Write down a differential equation for m.

(b) Given that the snowball has a mass of 12 units when $t = 0$, show that $m = 12\,e^{-kt}$, where k is a positive constant.

(c) Given that $m = 8$ units when $t = 5$, find the value of t at which the snowball is half its initial mass.

Review Advancing Maths for AQA C3C4 pages 300–305.

Test yourself ANSWERS

1 (a) $\dfrac{dy}{dx} = \sqrt{xy}$ **(b)** $\dfrac{dv}{dt} = p - kv^2$

2 $\tan x = t + \dfrac{1}{2}\sin 2t + c$

3 $y = 2xe^{\frac{1}{2}x^2 - 2}$

4 (a) $\dfrac{dm}{dt} = -km$, where k is a positive constant

(c) $t = \dfrac{5\ln 2}{\ln 1.5} = 8.55$ (to 3 sf)

CHAPTER 9
Vector equations of lines

Key points to remember

1 A vector is represented by bold type such as **v**, or by using an arrow above the letters such as $\overrightarrow{AB}$ to indicate the direction.

2 The magnitude of the vector $\overrightarrow{AB}$ is written as $|\overrightarrow{AB}|$ or as AB. The magnitude of the vector **v** is written as $|\mathbf{v}|$ or as v.

3 In two dimensions, when a vector **v** is written as $\mathbf{v} = \begin{bmatrix} a \\ b \end{bmatrix}$, the quantities a and b are the components of the vector in the x and y directions respectively.

4 In three dimensions, when a vector **v** is written as $\mathbf{v} = \begin{bmatrix} a \\ b \\ c \end{bmatrix}$, the quantities a, b and c are the components of the vector in the x, y and z directions respectively.

5 If the point A has position vector **a** and the point B has position vector **b**, then $\overrightarrow{AB} = \mathbf{b} - \mathbf{a}$.

6 In two dimensions, the magnitude of the vector $\begin{bmatrix} a \\ b \end{bmatrix}$ is $\sqrt{a^2 + b^2}$.

7 In three dimensions, the magnitude of the vector $\begin{bmatrix} a \\ b \\ c \end{bmatrix}$ is $\sqrt{a^2 + b^2 + c^2}$.

8 The distance between the points (x_1, y_1, z_1) and (x_2, y_2, z_2) is
$$\sqrt{(x_2 - x_1)^2 + (y_2 - y_1)^2 + (z_2 + z_1)^2}.$$

9 The scalar product of **p** and **q**, written as **p.q**, has value $|\mathbf{p}| \times |\mathbf{q}| \times \cos\theta$ or $pq\cos\theta$, where θ is the angle between the two vectors.

10 If $\mathbf{p} = \begin{bmatrix} a \\ b \\ c \end{bmatrix}$ and $\mathbf{q} = \begin{bmatrix} d \\ e \\ f \end{bmatrix}$ then **p.q** is evaluated as $\begin{bmatrix} a \\ b \\ c \end{bmatrix} \cdot \begin{bmatrix} d \\ e \\ f \end{bmatrix} = ad + be + cf.$

11 If $\mathbf{a}.\mathbf{b} = 0$, where **a** and **b** are non-zero vectors, then **a** and **b** are perpendicular.

12 In order to find the angle between two lines you find the angle between the direction vectors of the lines.

The acute angle, θ, between $\mathbf{r}_1 = \mathbf{a} + \lambda\mathbf{p}$ and $\mathbf{r}_2 = \mathbf{b} + \mu\mathbf{q}$ is given by $\cos\theta = \dfrac{|\mathbf{p}.\mathbf{q}|}{pq}$

Worked example 1

(a) Find the magnitude of the vector $\begin{bmatrix} -5 \\ 12 \end{bmatrix}$.

(b) The vectors $\begin{bmatrix} -5 \\ 12 \end{bmatrix}$ and $\begin{bmatrix} 3-p \\ p+1 \end{bmatrix}$ are perpendicular.

   Find the value of the constant p.

(a) The vector $\begin{bmatrix} -5 \\ 12 \end{bmatrix}$ has magnitude

$$\sqrt{(-5)^2 + 12^2} = \sqrt{25 + 144} = \sqrt{169} = 13.$$ — Using **6**

(b) Since the vectors are perpendicular, $\begin{bmatrix} -5 \\ 12 \end{bmatrix} \cdot \begin{bmatrix} 3-p \\ p+1 \end{bmatrix} = 0.$ — Using **11**

Hence $-5(3 - p) + 12(p + 1) = 0$ — Using **10**

$\Rightarrow -15 + 5p + 12p + 12 = 0$

$\Rightarrow 17p = 3 \quad \Rightarrow p = \dfrac{3}{17}$

Worked example 2

The triangle ABC has vertices $A(1, 2, 5)$, $B(3, -1, 4)$ and $C(2, -4, 6)$.

(a) Find the coordinates of the point D so that $ABCD$ is a parallelogram.

(b) Show that $9 + 14\cos ABC = 0$.

(c) (i) Find a vector equation of the line AB.

   (ii) The point $E(-5, 11, k)$ lies on the line AB. Find the value of the constant k.

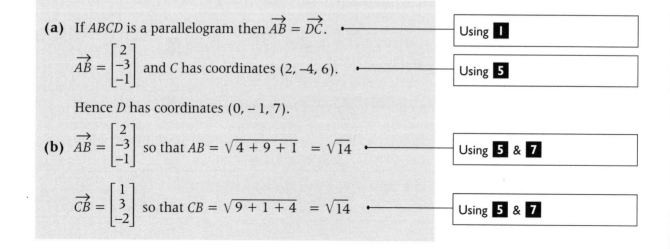

(a) If $ABCD$ is a parallelogram then $\overrightarrow{AB} = \overrightarrow{DC}$. — Using **1**

$\overrightarrow{AB} = \begin{bmatrix} 2 \\ -3 \\ -1 \end{bmatrix}$ and C has coordinates $(2, -4, 6)$. — Using **5**

Hence D has coordinates $(0, -1, 7)$.

(b) $\overrightarrow{AB} = \begin{bmatrix} 2 \\ -3 \\ -1 \end{bmatrix}$ so that $AB = \sqrt{4 + 9 + 1} = \sqrt{14}$ — Using **5** & **7**

$\overrightarrow{CB} = \begin{bmatrix} 1 \\ 3 \\ -2 \end{bmatrix}$ so that $CB = \sqrt{9 + 1 + 4} = \sqrt{14}$ — Using **5** & **7**

$$\vec{AB} \cdot \vec{CB} = \begin{bmatrix} 2 \\ -3 \\ -1 \end{bmatrix} \cdot \begin{bmatrix} 1 \\ 3 \\ -2 \end{bmatrix} = 2 - 9 + 2 = -9$$

Using **10**

$$\cos ABC = \frac{\vec{AB}.\vec{CB}}{AB \times CB} = \frac{-9}{\sqrt{14}\sqrt{14}} = \frac{-9}{14}$$

Using **12**

Hence $9 + 14\cos ABC = 0$.

(c) (i) The point $A(1, 2, 5)$ lies on the line AB and its direction is given by

$$\vec{AB} = \begin{bmatrix} 2 \\ -3 \\ -1 \end{bmatrix}.$$

A vector equation for the line AB is $\mathbf{r} = \begin{bmatrix} 1 \\ 2 \\ 5 \end{bmatrix} + \lambda \begin{bmatrix} 2 \\ -3 \\ -1 \end{bmatrix}$.

(ii) $\begin{bmatrix} -5 \\ 11 \\ k \end{bmatrix} = \begin{bmatrix} 1 \\ 2 \\ 5 \end{bmatrix} + \lambda \begin{bmatrix} 2 \\ -3 \\ -1 \end{bmatrix}$

The first two rows are consistent if $\lambda = -3$

since $-5 = 1 - 6$ and $11 = 2 + 6$

Hence $k = 5 - 3(-1) = 5 + 3 = 8$.

Worked example 3

Find, to the nearest 0.1°, the acute angle between the two lines with vector equations

$$\mathbf{r}_1 = \begin{bmatrix} 4 \\ -2 \\ 3 \end{bmatrix} + \lambda \begin{bmatrix} 3 \\ -1 \\ -2 \end{bmatrix} \text{ and } \mathbf{r}_2 = \begin{bmatrix} 0 \\ 5 \\ 7 \end{bmatrix} + \mu \begin{bmatrix} 2 \\ -6 \\ 5 \end{bmatrix}.$$

Note that when two lines are being considered, two different parameters are used such as λ and μ.

9

The angle, θ, between the two lines is equal to the angle between the two direction vectors $\begin{bmatrix} 3 \\ -1 \\ -2 \end{bmatrix}$ and $\begin{bmatrix} 2 \\ -6 \\ 5 \end{bmatrix}$.

Consider the scalar product $\begin{bmatrix} 3 \\ -1 \\ -2 \end{bmatrix} \cdot \begin{bmatrix} 2 \\ -6 \\ 5 \end{bmatrix} = 6 + 6 - 10 = 2$

Using **10**

Magnitude of $\begin{bmatrix} 3 \\ -1 \\ -2 \end{bmatrix} = \sqrt{9 + 1 + 4} = \sqrt{14}$

Using **7**

Magnitude of $\begin{bmatrix} 2 \\ -6 \\ -5 \end{bmatrix} = \sqrt{4 + 36 + 25} = \sqrt{65}$

Using **7**

Hence $\cos \theta = \dfrac{2}{\sqrt{14}\sqrt{65}} \approx 0.066299\ldots$

Using **12**

Hence the angle θ is 86.2° to the nearest 0.1°.

Worked example 4

(a) Find a vector equation of the line l_1 which passes through the points $A(1, 6, 4)$ and $B(5, 4, 7)$.

(b) Determine whether the line l_2 with equation

$$\mathbf{r} = \begin{bmatrix} 3 \\ -2 \\ 5 \end{bmatrix} + s \begin{bmatrix} 2 \\ 1 \\ -4 \end{bmatrix} \text{ intersects the line } l_1.$$

(a) The direction of the line is given by the vector

$$\overrightarrow{AB} = \begin{bmatrix} 4 \\ -2 \\ 3 \end{bmatrix}.$$

Using **5**

Hence a vector equation of the line l_1 is

$$\mathbf{r} = \begin{bmatrix} 1 \\ 6 \\ 4 \end{bmatrix} + t \begin{bmatrix} 4 \\ -2 \\ 3 \end{bmatrix}, \text{ where } t \text{ is a scalar parameter.}$$

(b) Lines would intersect when

$$\begin{bmatrix} 3 \\ -2 \\ 5 \end{bmatrix} + s \begin{bmatrix} 2 \\ 1 \\ -4 \end{bmatrix} = \begin{bmatrix} 1 \\ 6 \\ 4 \end{bmatrix} + t \begin{bmatrix} 4 \\ -2 \\ 3 \end{bmatrix}$$

Separating into 3 equations gives

$$3 + 2s = 1 + 4t$$
$$-2 + s = 6 - 2t$$
$$5 - 4s = 4 + 3t$$

The first two equations can be solved by doubling the second equation and adding the first:

$$-1 + 4s = 13 + 0 \Rightarrow 4s = 14 \Rightarrow s = \frac{7}{2}.$$

Substituting into the second equation gives

$$-2 + \frac{7}{2} = 6 - 2t \Rightarrow t = \frac{9}{4}.$$

However, the third equation

$$5 - 4s = 4 + 3t$$

is not satisfied by $s = \frac{7}{2}$ and $t = \frac{9}{4}$.

Hence the 2 lines do not intersect.

Worked example 5

The point A has coordinates $(2, -1, 3)$ and the line l has equation

$$\mathbf{r} = \begin{bmatrix} 2 \\ 5 \\ -1 \end{bmatrix} + \lambda \begin{bmatrix} 3 \\ -1 \\ 2 \end{bmatrix}.$$

(a) The point P on the line l is where $\lambda = p$. Show that

$$\overrightarrow{AP} \cdot \begin{bmatrix} 3 \\ -1 \\ 2 \end{bmatrix} = 14p - 14.$$

(b) Hence find the coordinates of the foot of the perpendicular from the point A to the line l.

(c) Determine the shortest distance from A to the line l.

(a) The point P has coordinates $(2 + 3p, 5 - p, -1 + 2p)$
Since A is the point $(2, -1, 3)$, the vector

$$\overrightarrow{AP} = \begin{bmatrix} 3p \\ 6 - p \\ -4 + 2p \end{bmatrix}.$$

> Using **5**

Hence $\overrightarrow{AP} \cdot \begin{bmatrix} 3 \\ -1 \\ 2 \end{bmatrix} = 3(3p) + (-1)(6 - p) + 2(-4 + 2p)$

> Using **10**

$$= 9p - 6 + p - 8 + 4p = 14p - 14.$$

(b) If P is the foot of the perpendicular from the point A to the line l, then AP is perpendicular to the line.

$$\overrightarrow{AP} \cdot \begin{bmatrix} 3 \\ -1 \\ 2 \end{bmatrix} = 0 \Rightarrow 14p - 14 = 0. \text{ Therefore } p = 1.$$

This gives the coordinates of P as $(5, 4, 1)$.

> Since P has coordinates $(2 + 3p, 5 - p, -1 + 2p)$

9

(c) The shortest distance from A to the line is the distance between $A(2, -1, 3)$ and $P(5, 4, 1)$.
$$AP^2 = 3^2 + 5^2 + (-2)^2 = 9 + 25 + 4 = 38.$$

> Using **8**

The shortest distance from A to the line l is $\sqrt{38}$.

REVISION EXERCISE 9

1 (a) Find the magnitude of the vector $\begin{bmatrix} -3 \\ 4 \end{bmatrix}$.

(b) The vectors $\begin{bmatrix} -3 \\ 4 \end{bmatrix}$ and $\begin{bmatrix} 2 + p \\ p - 3 \end{bmatrix}$ are perpendicular.
Find the value of the constant p.

2 Find the angle between the two vectors $\begin{bmatrix} 3 \\ 5 \end{bmatrix}$ and $\begin{bmatrix} 2 \\ -1 \end{bmatrix}$,
giving your answer to the nearest tenth of a degree.

3 The triangle PQR has vertices $P(1, 4, -5)$, $Q(2, 6, 1)$ and
$R(3, -3, 5)$.

(a) Determine which of the points P, Q or R is closest to the
origin.

(b) Show that the cosine of angle PQR is equal to $-\dfrac{1}{\sqrt{82}}$.

(c) The point $S(7, 3, k)$ is such that angle QRS is a right
angle. Find the value of the constant k.

4 Two lines have vector equations $\mathbf{r}_1 = \begin{bmatrix} 1 \\ -6 \\ 7 \end{bmatrix} + s \begin{bmatrix} -1 \\ 1 \\ -1 \end{bmatrix}$
and $\mathbf{r}_2 = \begin{bmatrix} 3 \\ -1 \\ 1 \end{bmatrix} + t \begin{bmatrix} 5 \\ 2 \\ -3 \end{bmatrix}$.

(a) Prove that the two lines intersect and find the
coordinates of the point of intersection.

(b) Find the angle between the two lines.

5 (a) Find a vector equation, with parameter t, of the line l_1
which passes through the points $A(2, -1, 3)$ and
$B(4, -3, 2)$.

(b) Determine whether the line l_2 with equation
$$\mathbf{r} = \begin{bmatrix} 3 \\ -2 \\ 5 \end{bmatrix} + s \begin{bmatrix} 2 \\ 1 \\ -2 \end{bmatrix}$$ intersects the line l_1.

(c) Show that the acute angle between the lines l_1 and l_2 is
$\cos^{-1}\left(\dfrac{4}{9}\right)$.

(d) The point C has coordinates $(3, 2, k)$. Find the value of k
so that angle ABC is a right angle.

6 In two dimensions, the point A has coordinates $(2, -11)$ and the point P lies on the line

$$\mathbf{r} = \begin{bmatrix} 4 \\ -1 \end{bmatrix} + \lambda \begin{bmatrix} 3 \\ 2 \end{bmatrix}, \text{ where } \lambda = p.$$

(a) Show that $\overrightarrow{AP}.\begin{bmatrix} 3 \\ 2 \end{bmatrix} = 26 + 13p.$

(b) Hence find the point on the line closest to A.

7 The points P and Q have coordinates $(-2, -2, 8)$ and $(2, -1, 4)$ respectively and O is the origin.

(a) Given that $\overrightarrow{OR} = 3\overrightarrow{OQ}$, find the coordinates of R.

(b) Show that PR has length 9.

(c) Find angle QPR, giving your answer to the nearest $0.1°$.

(d) The point $S(2, k, 5)$ is such that QS is perpendicular to PR. Find the value of k.

8 The points A and B have coordinates $(-1, 3, 1)$ and $(3, -2, 4)$ respectively.

The line l has equation $\mathbf{r} = \begin{bmatrix} 3 \\ -2 \\ 4 \end{bmatrix} + \lambda \begin{bmatrix} 1 \\ 2 \\ -1 \end{bmatrix}$

(a) Show that the distance between the points A and B is $5\sqrt{2}$.

(b) The line AB makes an acute angle θ with l. Find the value of θ, giving your answer to the nearest tenth of a degree.

(c) The point P on the line l is where $\lambda = p$. Show that

$$\overrightarrow{AP}.\begin{bmatrix} 1 \\ 2 \\ -1 \end{bmatrix} = 6p - 9.$$

(d) Hence find the coordinates of the foot of the perpendicular from the point A to the line l.

9 The point P has coordinates $(2, 2, -1)$ and the line l has

equation $\mathbf{r} = \begin{bmatrix} 4 \\ 3 \\ 1 \end{bmatrix} + \lambda \begin{bmatrix} -1 \\ 2 \\ 5 \end{bmatrix}.$

(a) The point Q on the line l is where $\lambda = q$. Show

that $\overrightarrow{PQ}.\begin{bmatrix} -1 \\ 2 \\ 5 \end{bmatrix} = 10 + 30q.$

(b) Hence find the coordinates of the foot of the perpendicular from the point P to the line l.

(c) Determine the shortest distance from P to the line l.

9

Test yourself	**What to review**
	If your answer is incorrect:

1 Find the magnitude of the vector $\begin{bmatrix} 2 \\ 1 \\ -2 \end{bmatrix}$.

Review Advancing Maths for AQA C3C4 pages 306–310.

---

2 Find the distance between the points $(-5, 3, 7)$ and $(-2, -2, 6)$.

Review Advancing Maths for AQA C3C4 pages 311–312.

---

3 Solve the equation $\begin{bmatrix} 2 \\ k \end{bmatrix} \cdot \begin{bmatrix} 3k \\ -2 \end{bmatrix} = 12$.

Review Advancing Maths for AQA C3C4 pages 312–315.

---

4 Find the angle between the two vectors

$\begin{bmatrix} 1 \\ 4 \\ -3 \end{bmatrix}$ and $\begin{bmatrix} 2 \\ -5 \\ -4 \end{bmatrix}$, giving your answer to the nearest $0.1°$.

Review Advancing Maths for AQA C3C4 pages 316–319.

---

5 Show that the line l_1 with equation

$\mathbf{r} = \begin{bmatrix} 1 \\ -2 \\ 4 \end{bmatrix} + s \begin{bmatrix} 3 \\ 2 \\ -1 \end{bmatrix}$ and the line l_2 which passes through

the points $(1, 2, 5)$ and $(-1, 2, 6)$ intersect. Find the position vector of their point of intersection.

Review Advancing Maths for AQA C3C4 pages 320–325.

---

6 The point A has coordinates $(3, 1, 5)$ and the line l has

equation $\mathbf{r} = \begin{bmatrix} -1 \\ 3 \\ 2 \end{bmatrix} + \lambda \begin{bmatrix} 1 \\ -1 \\ 4 \end{bmatrix}$.

Review Advancing Maths for AQA C3C4 pages 325–329.

(a) The point P on the line l is where $\lambda = p$. Show

that $\overrightarrow{AP} \cdot \begin{bmatrix} 1 \\ -1 \\ 4 \end{bmatrix} = 18p - 18$.

(b) Hence find the coordinates of the foot of the perpendicular from the point A to the line l.

(c) Determine the shortest distance from A to the line l.

Test yourself ANSWERS

1 3

2 $\sqrt{35}$

3 $k = 3$

4 $100.1°$

5 $\begin{bmatrix} 7 \\ 2 \\ 2 \end{bmatrix}$

6 (b) $(0, 2, 6)$ **(c)** $\sqrt{11}$

Exam style practice paper

Answer **all** questions.
Time allowed: 1 hour 30 minutes

1 A car depreciates according to the model $V = 24000e^{-kt}$,
   where £V is the value of the car t years after it is first
   sold as new.

   (a) State the value of the car when first sold as new.   (1 mark)
   (b) The value of the car exactly 1 year after it was
            first sold is £18000. Find the value of the car
            exactly 3 years after it was first sold.   (3 marks)

2 **(a)** Express $\sin x + 2\cos x$ in the form $R\sin(x + \alpha)$,

           where R is a positive constant and $0 < \alpha < \dfrac{\pi}{2}$.

           Give your value of α to two decimal places.   (3 marks)

   (b) The function f, with domain $0 < x < 4\pi$, is
            defined by $f(x) = \sin x + 2\cos x$.
            (i)   State the greatest value of $f(x)$.   (1 mark)
            (ii) Find the values of x for $0 < x < 4\pi$, at
                      which $f(x)$ takes its greatest value. Give
                      your answers to two significant figures.   (3 marks)

3 **(a)** When $6x^3 - x^2 - kx + 2$ is divided by $2x + 1$ the

           remainder is $3\dfrac{1}{2}$. Show that $k = 5$.   (2 marks)

   (b) The polynomial p(x) is defined by
            $p(x) = 6x^3 - x^2 - 5x + 2$.
            (i)   Use the factor theorem to show that $3x - 2$
                      is a factor of p(x).   (2 marks)
            (ii) Hence write $\dfrac{p(x)}{3x^2 + x - 2}$ in its simplest form.
                                                                  (4 marks)

4 (a) Find the binomial expansion of $(1 + 8x)^{\frac{1}{2}}$ up to the term in x^3, in its simplest form. (4 marks)

(b) State the range of values of x for which the expansion is valid. (1 mark)

(c) Use your answer to part **(a)** with $x = \dfrac{1}{100}$ to show that $\sqrt{3} \approx \dfrac{16238}{9375}$ (3 marks)

5 A curve is defined by the parametric equations

$$x = 2 + \frac{3}{t} \quad y = 1 + 2t.$$

(a) Find the coordinates of the point P on the curve where $t = \dfrac{1}{2}$. (2 marks)

(b) Find $\dfrac{dy}{dx}$ in terms of t. (4 marks)

(c) The tangent to the curve at P intersects the x-axis at the point A and the y-axis at the point B. Find the area of triangle OAB, where O is the origin. (5 marks)

(d) Show that the cartesian equation of the curve can be written as $xy - x - 2y = k$ and state the value of k. (2 marks)

6 (a) Solve the differential equation $\dfrac{dy}{dx} = y \cot x$,

given that $y = 3$ when $x = \dfrac{\pi}{2}$. (5 marks)

(b) Hence find the value of x in the interval

$0 < x < \dfrac{\pi}{2}$ when $y = 1.5$. (2 marks)

7 The points A and B have coordinates $(2, 3, -1)$ and $(3, 5, -3)$ respectively.

The line l has equation $\mathbf{r} = \begin{bmatrix} 1 \\ 2 \\ -3 \end{bmatrix} + \lambda \begin{bmatrix} 3 \\ 4 \\ 2 \end{bmatrix}$.

(a) Show that the distance between the points A and B is 3. (2 marks)

(b) Show that the line l intersects the line joining the points A and B and find the position vector of their point of intersection. (5 marks)

(c) Find the acute angle between l and the line AB. (4 marks)

8 (a) (i) Express $\dfrac{3x^2 - 6x}{(x+1)^2(2x-1)}$ in the form

$\dfrac{A}{x+1} + \dfrac{B}{(x+1)^2} + \dfrac{C}{2x-1}$, where

A, B and C are integers. (5 marks)

(ii) Hence find $\displaystyle\int \dfrac{3x^2 - 6x}{(x+1)^2(2x-1)}\ \mathrm{d}x.$ (4 marks)

(b) (i) Show that

$(\sin x + 3\cos x)^2 \equiv 5 + 3\sin 2x + 4\cos 2x.$ (4 marks)

(ii) Hence find $\displaystyle\int_{0}^{\frac{\pi}{2}} (\sin x + 3\cos x)^2\,\mathrm{d}x.$ (4 marks)

Answers

Revision exercise 1

1 **(a)** $1 + 3x + 9x^2 + 27x^3$ **(b)** 63

2 **(a)** $1 - 6x + 27x^2 - 108x^3$ **(b)** $|x| < \frac{1}{3}$

3 **(a)** **(i)** $1 - \frac{1}{3}x - \frac{1}{9}x^2 - \frac{5}{81}x^3$

        (ii) $4 + \frac{1}{8}x - \frac{1}{512}x^2 + \frac{1}{16384}x^3$

        (iii) $4 - 12x + 27x^2 - 54x^3$

    (b) **(i)** $|x| < 1$ **(ii)** $|x| < 16$ **(iii)** $|x| < \frac{2}{3}$

4 **(a)** $1 + \frac{1}{3}x^3 + \frac{2}{9}x^6 + \frac{14}{81}x^9$ **(b)** 0.30068

5 **(a)** $1 - 6x + 24x^2 - 80x^3$, $|x| < \frac{1}{2}$ **(b)** -280

6 **(a)** $1 - 4x + 10x^2$

    (b) **(i)** $1 - 6x + \frac{45}{2}x^2$

        (ii) $|x| < \frac{2}{3}$

    (c) $16 - 96x + 360x^2$

7 **(a)** $1 + \frac{1}{4}x - \frac{3}{32}x^2$

8 **(a)** $1 - \frac{1}{2}x + \frac{3}{8}x^2 - \frac{5}{16}x^3$ **(b)** $1 + \frac{5}{2}x + \frac{75}{8}x^2 + \frac{625}{16}x^3$

9 $a = -2$, coefficient $= 24$

10 **(b)** **(i)** $\frac{1}{2} + \frac{1}{4}x + \frac{1}{8}x^2 + \frac{1}{16}x^3$ **(ii)** $1 - 2x + 4x^2 - 8x^3$

    (c) $\frac{5}{2} - \frac{15}{4}x + \frac{65}{8}x^2 - \frac{255}{16}x^3$ **(d)** $|x| < \frac{1}{2}$

Revision exercise 2

1 (a) $\dfrac{2x}{x-1}$ (b) $\dfrac{x+2}{x-1}$ (c) $x+4$

2 $\dfrac{3}{x+1}$

3 (a) $-\dfrac{1}{x-3}$ (b) 1

4 (a) $3x^2 - 4x + 6$ (b) -15

5 (b) $(2x+1)(x-2)(x+2)$ (c) $\dfrac{(x-2)(x+2)}{x+4}$

6 (a) 2   (b) $A = -2, B = 5, C = 2$ (c) $9 + \ln\left(\dfrac{13}{7}\right)$

7 (a) -2 (b) -8

8 (b) $(3x-2)(2x^2 + x + 4)$

9 (b) $p = 2, q = 3$

   (c) $(2x+1)(x-1)(x-3)$ (d) $x = 0, x = \ln 3$

10 (a) $R = 10$ (b) $2x^2 + 5x + 8$ (c) $(2x-3)(2x^2 - 3x - 3)$

Revision exercise 3

1 (a) $2 - \dfrac{7}{x+1}$

2 (a) $2x + 1 + \dfrac{2}{2x-1}$ (b) $x^2 + x + \ln|2x-1| + c$

3 (a) $\dfrac{1}{x+2} + \dfrac{4}{x-1}$ (b) $\ln 20$

5 (a) $B = -2, C = 1.$

6 (a) $\dfrac{3}{3x+2} - \dfrac{2}{3x-2}$ (b) $\ln|3x+2| - \dfrac{2}{3}\ln|3x-2| + c$

7 (a) $A = 2, B = 2, C = -1$ (b) $3 - 4x - 2x^2 - 10x^3$

8 (a) $A = 3, B = -1, C = 2$ (b) $4 - 9x + 17x^2$

   (c) $-\dfrac{1}{2} < x < \dfrac{1}{2}$

9 (a) $3 - \dfrac{2}{2x+3} - \dfrac{4}{x+5}$ (b) $-16\dfrac{1}{8}$

10 (a) $2x - 3 - \dfrac{2}{x-1} + \dfrac{1}{4x+1}$ (b) 36

Revision exercise 4

1 (a) $1 + \dfrac{1}{y}\dfrac{dy}{dx}$ (b) $3x^2 + 3y^2\dfrac{dy}{dx} + \dfrac{dy}{dx} - 7$

   (c) $x(2\cos 2y\,\dfrac{dy}{dx}) + \sin 2y$ (d) $\dfrac{(y+1) - x\left(\dfrac{dy}{dx}\right)}{(y+1)^2}$

2 $-\dfrac{4}{5}$

3 $-\dfrac{3}{4}$

4 $\dfrac{dy}{dx} = \dfrac{2x\,(x^2 - y)^4}{\cos y + (x^2 - y)^4}$

5 $\dfrac{2}{3}$

6 (a) $\dfrac{dy}{dx} = \dfrac{4 - y}{x + 3y^2}$

7 (a) $\dfrac{dy}{dx} = 2y(1 + \ln x)$ (b) e^{-1}

8 (a) $A(-5, 0), B(3, 0)$ (d) $P\left(-1, -\dfrac{16}{3}\right)$

9 $(5, -1), (-1, 1)$

10 (a) $y = \dfrac{3}{2}x - 6$ (b) 12

11 (a) $(2, -9), (2, 3)$

   (b) $13x + 12y - 62 = 0, \quad 23x + 12y + 62 = 0$

12 (a) $(1, 0), (-1, 0)$

   (b) $(0, \dfrac{1}{2})$

13 (a) $\dfrac{dy}{dx} = \dfrac{2x + \cos y}{x \sin y - 2}$ (b) $\left(-\dfrac{1}{2}, \dfrac{25}{4}\right)$

14 (b) $(2, 2)$

15 (a) $\dfrac{dy}{dx} = \dfrac{3e^x\,(4e^{2x} - y^2)}{6ye^x - 1}$ (c) $a = -\ln 2, b = 1$

Revision exercise 5

1 (a) $(-9, 2), (-8, 1)$ (b) $(-5, 0), (0, -1), (0, 5)$

   (c) $x = y^2 - 4y - 5$

   (d)

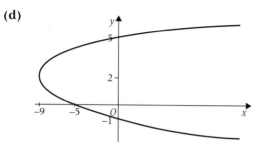

2 (a) $\dfrac{2t^2}{3}$ (b) $\dfrac{8}{3}$ (c) $3x + 8y = 47$

3 (a) (i) $2t, 3t^2$ (ii) $\dfrac{-3}{2}$ (iii) $3x + 2y = 11$

   (b) $(x - 4)^3 = (y + 1)^2$

4 (a) (i) $8t + 4$ (ii) $y = 12x - 52$ (b) $y = (x - 3)(x + 1)$

5 (a) (i) $-\left(\dfrac{3t^2 + 1}{2t}\right)$ (ii) $4x + 13y + 130 = 0$

6 (a) $(-2, 3)$ (b) $\dfrac{-3}{4}$ (c) $1, -3$

7 (a) (i) $x^3 = t^3 + 3t + \dfrac{3}{t} + \dfrac{1}{t^3}$ (ii) $x^3 = y + 3x$

   (b) (i) $1 - \dfrac{1}{t^2}, 3t^2 - \dfrac{3}{t^4}$ (ii) $\dfrac{21}{2}$

8 (a) (i) $\dfrac{1 - t^2}{t}$ (ii) $8y = 12x + 25$ (b) $27y^2 = 4(x + 1)(8 - x)^2$

9 (a) $\left(\dfrac{x - 3}{2}\right)^2 + (y - 4)^2 = 1$ (b) $\dfrac{1}{2}$

11 (a) (i) $t = 3 + \dfrac{1}{x}$ (b) (i) $\dfrac{-1}{(t - 3)^2}, \dfrac{-1}{(t + 1)^2}$ (ii) $\dfrac{1}{25}$

Revision exercise 6

1 (a) sin 50°   (b) cos 15°   (c) tan 62°

2 (a) sin 2x   (b) −90°, 0°, 90°, 180°

3 (a) cos 2x   (b) $2x + y = \dfrac{\pi}{2}$

4 (a) $x - \dfrac{1}{2}\sin 2x + c$ (b) $\dfrac{\pi - 3}{12}$

5 (a) 2sin (x + 30°)   (b) 2; −300°, 60°

6 (b) 0°, 36.9°, 180°, 323.1°, 360°

7 14°, 90°, 166°, 270°

8 (a) $\cos^2 x - \sin^2 x$ (b) 0°, 180°, 360°

9 (a) 13cos (x + 22.6°)   (b) −75.7°, 30.5°

10 (a) 10sin (x − 0.927)

    (b) (i) 10cos (x − 0.927)   (ii) 2.09, 6.05

11 (b) −75°, −15°, 105°, 165°

12 tan x − 4cos x + sin 2x + x + c

13 (b) (i)   sin 2x − x + c

14 (b) 15°, 75°, 105°, 165°

15 (a) $\tan 2x = \dfrac{2\tan x}{1 - \tan^2 x}$

    (c) 0, π, 2π, 0.293, 2.85, 3.43, 5.99

Revision exercise 7

1 (a) 3.696   (b) 0.471   (c) − 0.511

3 (a) £200   (b) £207.32   (c) 306 months

4 (a) 75°C   (b) 57.2°C   (c) 36.6 minutes

5 (b) 66 million

6 (b) 13.9

7 (b) 6250

8 1.02 pm

9 (a) $m = 7 \times 10^{-3} \times e^{-kt}$ where $k = \dfrac{1}{40}\ln\left(\dfrac{10}{9}\right)$ (b) 154 days

10 (a) $N = 700e^{kt}$ where $k = \dfrac{1}{20}\ln\left(\dfrac{30}{7}\right)$ (b) 32 days

11 (a) $P = 12e^{kt}$ million where $k = \dfrac{1}{6}\ln\left(\dfrac{7}{6}\right)$, 17.6 million

12 (a) $M = 500e^{kt}$ where $k = \dfrac{1}{8}\ln (1.24)$, 26 years

13 (a) £8000 (b) 7 years

Revision exercise 8

1 $\frac{y^2}{2} = \frac{x^2}{2} + 4x + c$

2 **(a)** $\frac{y^2}{2} = 2x - \frac{x^2}{2} + c$ **(b)** $r = \sqrt{8}$, centre $(2, 0)$

3 **(a)** $C = 2, D = -2$ **(b) (i)** $y = \frac{Ax^2}{(x+2)^2}$ **(ii)** 12.5

4 **(a)** $\frac{dV}{dt} = k\sqrt{V}, k > 0$ **(c)** 2012

5 **(a)** $\frac{1}{4}t = \ln\left(\frac{9}{8-y}\right)$ **(b)** 2.35 **(c)** $y = 8 - 9e^{-\frac{1}{4}t}$

   (d) 8

6 **(a)** $t \sin t + \cos t + c$ **(b)** $x = \frac{(\pi + 2)^2}{16}$

7 **(a)** $-\frac{1}{1 + \tan x} + c$

   (b) (i) $2\sqrt{y} = -\frac{1}{1 + \tan x} + 3$ **(ii)** $\frac{25}{16}$

8 **(a)** $\frac{dV}{dt} = R$, where V cm³ is the volume of water in the

   container at time t seconds.

   (b) (i) 24 **(iii)** $\frac{84}{\ln 2}$

9 **(b)** Minimum **(c)** $y^2 = x^2 + 2x + 5$

10 **(a) (i)** $\frac{1}{x+1} - \frac{1}{x+3}$ **(ii)** $\frac{x+1}{x+3} = Ae^t$

   (c) −5 **(d)** ln 1.6

11 **(b)** $r^3 = 5000[25 + 3\ln(1 + t)]$ **(c)** 59 m

Revision exercise 9

1 **(a)** 5 **(b)** $p = 18$

2 85.6°

3 **(a)** Q **(c)** 17.5

4 **(a)** $(-2, -3, 4)$ **(b)** 90°

5 **(a)** $\mathbf{r} = \begin{bmatrix} 2 \\ -1 \\ 3 \end{bmatrix} + t\begin{bmatrix} 2 \\ -2 \\ -1 \end{bmatrix}$ **(b)** do not intersect **(d)** −10

6 $(-2, -5)$

7 **(a)** $R(6, -3, 12)$ **(c)** 73.1° **(d)** $k = 3$

8 **(b)** 121.3° **(d)** $(4.5, 1, 2.5)$

9 **(b)** $\left(4\frac{1}{3}, 2\frac{1}{3}, -\frac{2}{3}\right)$ **(c)** $\frac{\sqrt{51}}{3}$

Exam style practice paper

1 **(a)** £24000 **(b)** £10125

2 **(a)** $\sqrt{5}\sin(x+1.11)$ **(b) (i)** $\sqrt{5}$ **(ii)** 0.46, 6.7

3 **(b)** **(ii)** $2x-1$

4 **(a)** $1+4x-8x^2+32x^3$ **(b)** $|x|<\dfrac{1}{8}$

5 **(a)** $(8, 2)$ **(b)** $-\dfrac{2t^2}{3}$ **(c)** $\dfrac{100}{3}$ **(d)** $k=4$

6 **(a)** $y=3\sin x$ **(b)** $x=\dfrac{\pi}{6}$

7 **(b)** $\begin{bmatrix} 2.5 \\ 4 \\ -2 \end{bmatrix}$ **(c)** $64.3°$ (to 3 s.f.)

8 **(a)** **(i)** $\dfrac{2}{x+1}-\dfrac{3}{(x+1)^2}-\dfrac{1}{2x-1}$

(ii) $2\ln|x+1|+\dfrac{3}{x+1}-\dfrac{1}{2}\ln|2x-1|+c$

(b) **(ii)** $3+\dfrac{5\pi}{2}$